Rinaldo B. Schinazi

Classical and Spatial Stochastic Processes

With Applications to Biology

Second Edition

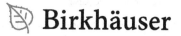 Birkhäuser

Rinaldo B. Schinazi
Department of Mathematics
University of Colorado
Colorado Springs, CO, USA

ISBN 978-1-4939-3710-3   ISBN 978-1-4939-1869-0 (eBook)
DOI 10.1007/978-1-4939-1869-0
Springer New York Heidelberg Dordrecht London

Mathematics Subject Classification (2010): 60-01, 60K35, 92-01

Preface to the Second Edition

The first part of this book concerns classical stochastic models and the second part concerns spatial stochastic models. For this second edition, the classical part of the book has been completely rewritten. The most important models, such as random walks, branching processes, and the Poisson process, are treated in separate chapters. For both discrete and continuous time models, we first study interesting examples and then we look at the general theory. Another important difference with the first edition is that most models are applied to population biology questions. We use stochastic models to test theoretical biology questions such as "Can the immune system be overcome by a fast mutating virus?" or "How far away is the next influenza pandemic?"

The second part of the book has been updated and streamlined. We cover the same spatial models as in the first edition. It is a small introduction to spatial stochastic models. We show that interesting results can be obtained with relatively elementary mathematics.

We also have added a first chapter that reviews important probability tools. The reader, however, is supposed to have had a first probability course. Many computations require solid Calculus (series in particular) and some proofs require knowing classical analysis. The appendix covers a few advanced probability results that are needed for the spatial models part. I have tried to make the chapters as independent as possible. At the beginning of each new chapter, there are some words about what is needed from previous chapters.

Acknowledgements. I would like to thank the colleagues who had kind words for the first edition of this book and, in particular, Norio Konno for translating the book into Japanese. I also thank Claude Bélisle for pointing out several typos and Bruno Monte for sharing his Latex expertise.

Colorado Springs, CO, USA   Rinaldo B. Schinazi

Preface to the First Edition

This book is intended as a text for a first course in stochastic processes at upper undergraduate or graduate levels, assuming only that the reader has had a serious Calculus course—advanced Calculus would even be better—as well as a first course in probability (without measure theory). In guiding the student from the simplest classical models to some of the spatial models, currently the object of considerable research, the text is aimed at a broad audience of students in biology, engineering, mathematics, and physics.

The first two chapters deal with discrete Markov chains—recurrence and transience, random walks, birth and death chains, ruin problem, and branching processes—and their stationary distributions. These classical topics are treated with a modern twist: in particular, the coupling technique is introduced in the first chapter and used throughout. The third chapter deals with continuous time Markov chains—Poisson process, queues, birth and death chains, stationary distributions.

The second half of the book treats spatial processes. This is the main difference between this work and the many others on stochastic processes. Spatial stochastic processes are (rightly) known to be difficult to analyze. The few existing books on the subject are technically challenging and intended for a mathematically sophisticated reader. We picked several interesting models—percolation, cellular automata, branching random walks, contact process on a tree—and concentrated on those properties that can be analyzed using elementary methods. These methods include contour arguments (for percolation and cellular automata) and coupling techniques (for branching random walks and the contact process). Even though this is only a small introduction to the subject, it deals with some very important and interesting properties of spatial stochastic processes, such as the existence of phase transitions, the continuity of phase transitions, and the existence of multiple critical points.

Colorado Springs, CO, USA   Rinaldo B. Schinazi

Contents

Chapter 1
A Short Probability Review

In this chapter we will review probability techniques that will be useful in the sequel.

1 Sample Space and Random Variables

1.1 Sample Space and Probability Axioms

The study of probability is concerned with the mathematical analysis of random experiments such as tossing a coin, rolling a die, or playing at the lottery. Each time we perform a random experiment there are a number of possible outcomes. The sample space Ω of a random experiment is the collection of all possible outcomes.

An event is any subset of Ω. A sequence of events A_i is said to be disjoint if for $i \neq j$ we have

$$A_i \cap A_j = \emptyset.$$

Example 1.1. Roll a die. The sample space is $\Omega = \{1, 2, 3, 4, 5, 6\}$. The event $B = \{1, 3, 5\}$ is the same as the event "the die showed an odd face."

Example 1.2. We count the number of rolls until we get a 6. Here $\Omega = \{1, 2, \dots\}$. The sample space consists of all strictly positive integers. Note that this sample space has infinitely many outcomes.

Definition 1.1. A probability is a function with the following properties.

1. The probability $P(A)$ of an event A is in $[0, 1]$.
2. The probability of the whole sample space $P(\Omega)$ is 1.
3. For a finite or infinite sequence of disjoint events A_i we have

$$P(\bigcup_i A_i) = \sum_i P(A_i).$$

© Springer Science+Business Media New York 2014
R.B. Schinazi, *Classical and Spatial Stochastic Processes: With Applications to Biology*, DOI 10.1007/978-1-4939-1869-0_1

1.2   Discrete Random Variables

A discrete random variable is a function from a sample space Ω into a countable set (usually the positive integers). The distribution of a random variable X is the sequence of probabilities $P(X = k)$ for all k in the range of X. We must have

$$P(X = k) \geq 0 \text{ for every } k \text{ and } \sum_k P(X = k) = 1.$$

1.2.1   Expectation and Variance

The expected value (or mean) of a discrete random variable is denoted by $E(X)$. The definition is

$$E(X) = \sum_k k P(X = k),$$

where the sum is on all possible values that X takes. More generally, for a function g we have

$$E(g(X)) = \sum_k g(k) P(X = k).$$

Note that the expected value need not exist (the corresponding series may diverge). An important particular case of the preceding formula is $g(x) = x^2$ for which we get

$$E(X^2) = \sum_k k^2 P(X = k).$$

The following property is quite important. Let X and Y be two random variables then

$$E(X + Y) = E(X) + E(Y).$$

We define the variance of a random variable by

$$Var(X) = E(X^2) - E(X)^2.$$

The expected value is a measure of location of X while the variance is a measure of dispersion of X.

We now give two examples of important discrete random variables.

1.2.2 Bernoulli Random Variables

These are the simplest possible random variables. Perform a random experiment with two possible outcomes: success or failure. Set $X = 1$ if the experiment is a success and $X = 0$ if the experiment is a failure. Such a 0–1 random variable is called a Bernoulli random variable. The usual notation is $P(X = 1) = p$ and $P(X = 0) = q = 1 - p$. Moreover,

$$E(X) = 1 \times p + 0 \times (1 - p) = p,$$

and

$$E(X^2) = 1^2 \times p + 0^2 \times (1 - p) = p.$$

Hence, $E(X) = p$ and $Var(X) = p - p^2 = pq$.

Example 1.2. Roll a fair die. We define success as rolling a 6. Thus, the probability of success is $P(X = 1) = 1/6$. We have $p = 1/6$ and $q = 5/6$.

1.2.3 Poisson Random Variables

The random variable N is said to have a Poisson distribution with parameter λ if

$$P(N = k) = e^{-\lambda}\frac{\lambda^k}{k!} \text{ for } k = 0, 1, \ldots.$$

As we will see below λ is also the mean of N.

The Poisson distribution is a good model for counting the number of occurrences of events that have small probabilities and are independent.

Example 1.3. Consider a fire station that serves a given neighborhood. Each resident has a small probability of needing help on a given day and most of the time people need help independently of each other. The number of calls a fire station gets on a given day may be modeled by a Poisson random variable with mean λ. Assume that $\lambda = 6$. What is the probability that a fire station get 2 or more calls in a given day?

$$P(N \geq 2) = 1 - P(N = 0) - P(N = 1) = 1 - e^{-\lambda} - \lambda e^{-\lambda} = 1 - 7e^{-6} \sim 0.98.$$

Recall that

$$e^x = \sum_{k=0}^{\infty} \frac{x^k}{k!} \text{ for every } x.$$

Hence,

$$\sum_{k=0}^{\infty} P(N = k) = \sum_{k=0}^{\infty} e^{-\lambda} \frac{\lambda^k}{k!} = e^{-\lambda} e^{\lambda} = 1.$$

This shows that the Poisson distribution is indeed a probability distribution.

We are now going to compute the mean of a Poisson random variable N with mean λ. We have

$$E(N) = \sum_{k=0}^{\infty} k P(N = k) = \sum_{k=1}^{\infty} k e^{-\lambda} \frac{\lambda^k}{k!} = e^{-\lambda} \lambda \sum_{k=1}^{\infty} \frac{\lambda^{k-1}}{(k-1)!}.$$

By shifting the summation index we get

$$\sum_{k=1}^{\infty} \frac{\lambda^{k-1}}{(k-1)!} = \sum_{k=0}^{\infty} \frac{\lambda^k}{k!} = e^{\lambda}.$$

Thus,

$$E(N) = e^{-\lambda} \lambda \sum_{k=1}^{\infty} \frac{\lambda^{k-1}}{(k-1)!} = e^{-\lambda} \lambda e^{\lambda} = \lambda.$$

2 Independence

Definition 2.1. Two events A and B are said to be independent if

$$P(A \cap B) = P(A)P(B).$$

Two discrete random variables X and Y are said to be independent if for all possible values k and m we have

$$P(\{X = k\} \cap \{Y = m\}) = P(X = k)P(Y = m).$$

Instead of the notation $P(\{X = k\} \cap \{Y = m\})$ we will often use the notation $P(X = k, Y = m)$. Intuitively, the event A is independent of the event B if knowing that B happened does not change the probability of A happening. This will be made clear with the notion of conditional probability that we will see below.

2.1 Binomial Random Variables

Consider n independent and identically distributed Bernoulli random variables $X_1, X_2 \ldots X_n$. Let p be the probability of success. That is, for $i = 1 \ldots n$

$$P(X_i = 1) = p.$$

Let B be the number of successes among these n experiments. Since trial i is a success if $X_i = 1$ and a failure if $X_i = 0$ we have

$$B = X_1 + X_2 + \cdots + X_n.$$

The random variable B is called a binomial distribution with parameters n and p. We now compute $P(B = k)$ for $k = 0, 1 \ldots, n$.

We first do an example. Assume $n = 3$ and $k = 2$. That is, we have two successes and one failure. There are three ways to achieve that. We may have SSF (first two trials are successes, the third one is a failure), SFS or FSS where S stands for success and F for failure. By independence, each possibility has probability $p^2 q$. Moreover, the three possibilities are disjoint. Hence,

$$P(B = 2) = 3p^2 q.$$

More generally, the number of possibilities for k successes in n trials is

$$\binom{n}{k} = \frac{n!}{k!(n-k)!}.$$

Each possibility has probability $p^k (1 - p)^{n-k}$ and the different possibilities are disjoint. Hence, for $k = 0, 1, \ldots, n$ we have the binomial distribution

$$P(B = k) = \binom{n}{k} p^k (1 - p)^{n-k}.$$

We can easily check that this is indeed a probability distribution by using the binomial theorem

$$(a + b)^n = \sum_{k=0}^{n} \binom{n}{k} a^k b^{n-k}.$$

Hence,

$$\sum_{k=0}^{n} P(B = k) = \sum_{k=0}^{n} \binom{n}{k} p^k (1 - p)^{n-k} = (p + 1 - p)^n = 1.$$

We now turn to the expected value. By using the linearity of the expectation we get

$$E(B) = E(X_1 + X_2 + \cdots + X_n) = E(X_1) + E(X_2) + \cdots + E(X_n) = np.$$

2.2   Geometric Random Variables

Example 2.1. Roll a fair die until you get a 6. Assume that the rolls are independent of each other. Let X be the number of rolls to get the first 6. The possible values of X are all strictly positive integers. Note that $X = 1$ if and only if the first roll is a 6. So $P(X = 1) = 1/6$. In order to have $X = 2$ the first roll must be anything but 6 and the second one must be 6. By independence of the different rolls we get $P(X = 2) = 5/6 \times 1/6$. More generally, in order to have $X = k$ the first $k - 1$ rolls cannot yield any 6 and the kth roll must be a 6. Thus,

$$P(X = k) = (5/6)^{k-1} \times 1/6 \text{ for all } k \geq 1.$$

Such a random variable is called geometric.

More generally, we have the following. Consider a sequence of independent identical trials. Assume that each trial can result in a success or a failure. Each trial has a probability p of success and $q = 1 - p$ of failure. Let X be the number of trials up to and including the first success. Then X is called a geometric random variable. The distribution of X is given by

$$P(X = k) = q^{k-1}p \text{ for all } k \geq 1.$$

Note that a geometric random variable may be arbitrarily large since the above probabilities are never 0. In order to check that the sum of these probabilities is 1 we need the following fact about geometric series.

$$\sum_{k \geq 0} x^k = \frac{1}{1 - x} \text{ for all } x \in (-1, 1).$$

Hence,

$$\sum_{k \geq 1} P(X = k) = \sum_{k \geq 1} q^{k-1}p = p \sum_{k \geq 0} q^k = \frac{p}{1 - q} = 1.$$

By taking the derivative of the geometric sum we get for x in $(-1, 1)$

$$\sum_{k \geq 1} kx^{k-1} = \frac{1}{(1 - x)^2}.$$

Applying this formula we get

$$E(X) = \sum_{k \geq 1} k q^{k-1} p = p \frac{1}{(1-q)^2} = \frac{1}{p}.$$

Not unexpectedly we see that the less likely the success the larger the expected value.

2.3 A Sum of Poisson Random Variables

Assume that N_1 and N_2 are independent Poisson random variables with parameters λ_1 and λ_2, respectively. What is the distribution of $N = N_1 + N_2$?

Let $n \geq 0$. We have

$$\{N = n\} = \bigcup_{k=0}^{n} \{N_1 = k, N_2 = n - k\}.$$

This is so because if $N = n$ then N_1 must be some $k \leq n$. If $N_1 = k$ then $N_2 = n - k$. Moreover, the events $\{N_1 = k, N_2 = n - k\}$ for $k = 0, \ldots n$ are disjoint (why?). Hence,

$$P(N = n) = \sum_{k=0}^{n} P(N_1 = k, N_2 = n - k).$$

Since N_1 and N_2 are independent we have for every k

$$P(N_1 = k, N_2 = n - k) = P(N_1 = k)P(N_2 = n - k).$$

Therefore

$$P(N = n) = \sum_{k=0}^{n} P(N_1 = k)P(N_2 = n - k).$$

Up to this point we have only used the independence assumption. We now use that N_1 and N_2 are Poisson distributed to get

$$P(N = n) = \sum_{k=0}^{n} e^{-\lambda_1} \frac{\lambda_1^k}{k!} e^{-\lambda_2} \frac{\lambda_2^{n-k}}{(n-k)!}.$$

By dividing and multiplying by $n!$ we get

$$P(N = n) = \frac{1}{n!} e^{-\lambda_1 - \lambda_2} \sum_{k=0}^{n} \binom{n}{k} \lambda_1^k \lambda_2^{n-k} = \frac{1}{n!} e^{-\lambda_1 - \lambda_2} (\lambda_1 + \lambda_2)^n,$$

where the last equality comes from the binomial theorem. This computation proves that $N = N_1 + N_2$ is also a Poisson distribution. It has rate $\lambda_1 + \lambda_2$.

3 Conditioning

Conditioning on the event B means restricting the sample space to B. The possible outcomes are then in B (instead of all of Ω). We define conditional probabilities next.

Definition 3.1. The probability of A given B is defined by

$$P(A|B) = \frac{P(A \cap B)}{P(B)},$$

assuming $P(B) > 0$.

The formula above is also useful in the form

$$P(A \cap B) = P(A|B)P(B).$$

Proposition 3.1. *The events A and B are independent if and only if*

$$P(A|B) = P(A).$$

In words, the probability of A does not change by conditioning on B. Knowing that B occurred does not add any information about A occurring or not. The proof of this result is easy and is left as an exercise.

The following rule is called the Rule of averages. It will turn out to be quite useful in a number of situations.

Proposition 3.2. *Assume that the events B_1, B_2, ... are disjoint and that their union is the whole sample space Ω. For any event A we have*

$$P(A) = \sum_i P(A|B_i)P(B_i),$$

where the sum can be finite or infinite.

To prove the formula note that the events $A \cap B_i$ for $i = 1, 2, \ldots$ are disjoint (why?) and that their union is A (why?). By the probability rules we have

$$P(A) = \sum_i P(A \cap B_i).$$

By the definition of conditional probability we have for every i

$$P(A \cap B_i) = P(A|B_i)P(B_i).$$

Hence,

$$P(A) = \sum_i P(A|B_i)P(B_i),$$

and this proves the formula.

Example 3.1. We have three boxes labeled 1, 2, and 3. Box 1 has 1 white ball and 2 black balls, Box 2 has 2 white balls and 1 black ball, and Box 3 has 3 white balls. One of the three boxes is picked at random and then a ball is picked (also at random) from this box. For $i = 1, 2, 3$ let A_i be the event "Box i is picked" and let B be the event "a white ball is picked." Are A_1 and B independent?

Given that the boxes are picked at random (uniformly) we have $P(A_1) = 1/3$. To compute $P(B)$ we use the rule of averages

$$P(B) = P(B|A_1)P(A_1) + P(B|A_2)P(A_2) + P(B|A_3)P(A_3).$$

We have $P(B|A_1) = 1/3$, $P(B|A_2) = 2/3$ and $P(B|A_3) = 3/3$. Hence,

$$P(B) = \frac{1}{3}(\frac{1}{3} + \frac{2}{3} + \frac{3}{3}) = \frac{2}{3}.$$

Therefore, $P(B|A_1)$ is not equal to $P(B)$. The events A_1 and B are not independent.

3.1 Thinning a Poisson Distribution

This is another application of the rule of averages.

Assume that N has a Poisson distribution with rate λ. Think of N as being the (random) number of customers arriving at a business during 1 h. Assume that customers arrive independently of each other. Assume also that the proportion of female customers is p. Let N_1 be the number of arriving female customers. The random variable N_1 is said to be a thinning of N (we only count the female customers). We will show now that N_1 is also a Poisson random variable.

Observe that given $N = n$ we can think of each arriving customer as being a failure (male customer) or success (female customer). Since we are assuming independence of arrivals and identical probability of success p we see that N_1 follows a binomial with parameters n and p. That is, for $k = 0, 1 \ldots n$ we have

$$P(N_1 = k|N = n) = \binom{n}{k} p^k (1 - p)^{n-k}.$$

By the rule of averages we have

$$P(N_1 = k) = \sum_{n=k}^{\infty} P(N_1 = k|N = n)P(N = n).$$

We now use the distribution of N and the conditional distribution of N_1 to get

$$P(N_1 = k) = \sum_{n=k}^{\infty} \binom{n}{k} p^k (1 - p)^{n-k} e^{-\lambda} \frac{\lambda^n}{n!}.$$

Expressing the binomial coefficient in terms of factorials

$$P(N_1 = k) = \frac{1}{k!} p^k \lambda^k e^{-\lambda} \sum_{n=k}^{\infty} \frac{1}{(n-k)!} (1-p)^{n-k} \lambda^{n-k}.$$

By a shift of index

$$\sum_{n=k}^{\infty} \frac{1}{(n-k)!} (1-p)^{n-k} \lambda^{n-k} = \sum_{n=0}^{\infty} \frac{1}{n!} (1-p)^n \lambda^n = e^{\lambda(1-p)}.$$

Therefore,

$$P(N_1 = k) = \frac{1}{k!} p^k \lambda^k e^{-\lambda} e^{\lambda(1-p)} = e^{-\lambda p} \frac{(\lambda p)^k}{k!}.$$

In words, the thinned random variable N_1 is also a Poisson random variable. Its rate is λp.

4 Generating Functions

We conclude this quick overview with the notion of generating function.

Definition 4.1. Let X be a discrete random variable whose values are positive integers. The generating function of a random variable X is defined by

$$g_X(s) = E(s^X) = \sum_{n \geq 0} s^n P(X = n).$$

Note that g_X is a power series. We have

$$|s^n P(X = n)| \leq |s|^n$$

and since the geometric series $\sum_{n \geq 0} |s|^n$ converges for $|s| < 1$ so does $\sum_{n \geq 0} s^n P(X = n)$ (by the comparison test). In other words, the generating function g_X is defined for s in $(-1, 1)$. It is also defined at $s = 1$ (why?).

Example 4.1. What is the generating function of a Poisson random variable N with rate λ?

By definition

$$g_N(s) = \sum_{n \geq 0} P(N = n) s^n = \sum_{n \geq 0} e^{-\lambda} \frac{\lambda^n}{n!} s^n.$$

Summing the series yields

$$g_N(s) = e^{-\lambda} e^{\lambda s} = e^{\lambda(-1+s)}.$$

The next result shows that there is one-to-one correspondence between generating functions and probability distributions.

Proposition 4.1. *Let X and Y be random variables with generating functions g_X and g_Y, respectively. Assume that for $|s| < 1$*

$$g_X(s) = g_Y(s).$$

Then, X and Y have the same distribution.

This is a direct consequence of the uniqueness of power series expansions. More precisely, if there is $R > 0$ such that for $|s| < R$ we have

$$\sum_{n\geq0} a_n s^n = \sum_{n\geq0} b_n s^n$$

then $a_n = b_n$ for all $n \geq 0$. For a proof see Rudin (1976), for instance.

Example 4.2. Assume that a random variable X has a generating function

$$g_X(s) = \frac{ps}{1 - qs}$$

where p is in $(0, 1)$ and $q = 1 - p$. What is the distribution of X?
  We use the geometric series to get

$$\frac{1}{1 - qs} = \sum_{n\geq0}(qs)^n$$

and therefore

$$g_X(s) = ps \sum_{n\geq0}(qs)^n = \sum_{n\geq0} pq^n s^{n+1} = \sum_{n\geq1} pq^{n-1} s^n,$$

where the last equality comes from a shift of index. Let Y be a geometric random variable with parameter p. We have for $n \geq 1$

$$P(Y = n) = pq^{n-1}$$

and therefore

$$g_Y(s) = \sum_{n\geq1} pq^{n-1} s^n.$$

Hence, $g_Y = g_X$ and by Proposition 4.1 X is also a geometric random variable with parameter p.

  The following property is quite useful.

Proposition 4.2. *If X and Y are two independent random variables, then*

$$g_{X+Y}(s) = g_X(s)g_Y(s).$$

Proof of Proposition 4.2.

$$g_{X+Y}(s) = \sum_{n \geq 0} s^n P(X + Y = n)$$

but

$$P(X + Y = n) = \sum_{k=0}^{n} P(X = k; Y = n - k) = \sum_{k=0}^{n} P(X = k)P(Y = n - k)$$

where the last equality comes from the independence of X and Y.

$$g_{X+Y}(s) = \sum_{n \geq 0} s^n \sum_{k=0}^{n} P(X = k)P(Y = n - k) = \sum_{n \geq 0} s^n P(X = n) \sum_{n \geq 0} s^n P(Y = n)$$

where the last equality comes from results about the product of two absolute convergent series, see Rudin (1976) for instance. This completes the proof of Proposition 4.2.

4.1   Sum of Poisson Random Variables (Again)

Assume that X and Y are two independent Poisson random variables with rates λ and μ, respectively. What is the distribution of $X + Y$?

We start with the generating function of X:

$$g_X(s) = \sum_{k \geq 0} s^k e^{-\lambda} \frac{\lambda^k}{k!} = e^{\lambda(s-1)}.$$

By Proposition 4.2 and Example 4.1 we have

$$g_{X+Y}(s) = g_X(s)g_Y(s) = e^{\lambda(s-1)}e^{\mu(s-1)} = e^{(\lambda+\mu)(s-1)}.$$

This is the generating function of a Poisson distribution with rate $\lambda + \mu$. Thus, by Proposition 4.1, $X + Y$ has a Poisson distribution with rate $\lambda + \mu$.

Note that we have already solved this problem in Sect. 2.3 by a different method. Using generating functions is easier and faster.

4.2 *Thinning a Poisson Distribution (Again)*

Assume that N has a Poisson distribution with rate λ. If $N = 0$, then let $N_1 = 0$. Assume $n \geq 1$. Given $N = n$ let the conditional distribution of N_1 be a binomial with parameters n and p. Hence, given $N = n$ we can write N_1 as the sum

$$N_1 = \sum_{i=1}^{n} X_i$$

where the random variables are independent Bernoulli random variables with distribution $P(X_i = 1) = p$ and $P(X_i = 0) = 1 - p$.

We now compute the generating function of N_1.

$$g_{N_1}(s) = E(s^{N_1}) = \sum_{n \geq 0} E(s^{N_1}|N = n)P(N = n),$$

where we are using a rule of averages for expectations. The conditional distribution of N_1 is a sum of independent Bernoulli random variables. Hence,

$$E(s^{N_1}|N = n) = E(s^{\sum_{i=1}^{n} X_i}) = E(s^X)^n.$$

Therefore,

$$g_{N_1}(s) = \sum_{n \geq 0} E(s^X)^n P(N = n) = g_N(E(s^X)).$$

Since N is a Poisson random variable with rate λ we have $g_N(s) = e^{\lambda(-1+s)}$. Since X is a Bernoulli random variable we have $E(s^X) = 1 - p + ps$. Hence,

$$g_{N_1}(s) = g_N(E(s^X)) = \exp(\lambda(-1 + 1 - p + ps)) = \exp(\lambda p(-1 + s)).$$

This proves that N_1 is a Poisson distribution with rate λp. Here again using generating functions simplifies the computations.

Problems

1. Show that the probability rules imply the following.

(a) If $A \cap B = \emptyset$, then

$$P(A \cup B) = P(A) + P(B).$$

(b) If A^c is the complement of A (everything not in A), then $P(A^c) = 1 - P(A)$.

(c) $P(\emptyset) = 0$. (Observe that $\Omega^c = \emptyset$.)

(d) For any events A and B we have

$$P(A) = P(AB) + P(AB^c).$$

(e) If $A \subset B$, then $P(A^c B) = P(B) - P(A)$.

2. Let N be a Poisson random variable with rate λ.

(a) Show that

$$E(N(N-1)) = \lambda^2.$$

(b) Use (a) to show that

$$E(N^2) = \lambda^2 + \lambda.$$

(c) Show that

$$Var(N) = \lambda.$$

3. Assume that the events $B_1, B_2, \ldots$ are disjoint and if that their union is the whole sample space Ω. Let A be an event.

(a) Let $C_i = A \cap B_i$ for $i = 1, 2, \ldots$. Show that the sets C_i are disjoint.

(b) Show that the union of the sets C_i is A.

4. Prove Proposition 3.1.

5. Let A and B two events such that $P(A) > 0$ and $P(B) > 0$. Show that A and B are disjoint if and only if they are not independent.

6. Let X be a geometric random variable with parameter p and let $q = 1 - p$.

(a) Show that for $k \geq 1$

$$P(X > k) = q^k.$$

(b) Show that

$$P(X > r + s | X > r) = P(X > s).$$

(c) Why is the property in (b) called the memoryless property of the geometric distribution?

7. (a) What is the generating function of a Bernoulli random variable with parameter p?

(b) Use (a) to show that the generating function of a binomial random variable with parameters n and p is

$$(1 - p + ps)^n.$$

8. Assume that X and Y are independent binomial random variables with parameters (n, p) and (m, p), respectively. Show that $X + Y$ is a binomial random variables with parameters $(n + m, p)$. (Use problem 7.)

9. The generating function of the random variable X is

$$g_X(s) = \frac{1}{n+1} \frac{1 - s^{n+1}}{1 - s},$$

where n is a fixed natural.

(a) Show that for $k = 0, 1, \ldots, n$

$$P(X = k) = \frac{1}{n+1}.$$

(Recall that $1 + s + \cdots + s^n = \frac{1-s^{n+1}}{1-s}$.)
(b) What is the distribution of X called?

10. Let $N \geq 0$ be a discrete random variable. Given $N = n$ we define

$$Y = \sum_{i=1}^{n} X_i$$

where $X_1, X_2, \ldots$ are independent and identically distributed random variables. Show the following relation between the different generating functions

$$g_Y(s) = g_N(g_X(s)).$$

Notes. There are many introductory texts in probability, see Schinazi (2011) for instance. Port (1994) is more advanced. It is neatly divided in self-contained short chapters and is a great reference for a number of topics. Rudin (1976) is a classic in analysis.

References

Port, S.C.: Theoretical Probability for Applications. Wiley, New York (1994)
Rudin, W.: Principles of Mathematical Analysis, 3rd edn. McGraw-Hill, Singapoore (1976)
Schinazi, R.B.: Probability with Statistical Applications, 2nd edn. Birkhauser, Basel (2011)

Chapter 2
Discrete Time Branching Process

We introduce branching processes. They are a recurring theme throughout the book. In this chapter we use them to model drug resistance and cancer risk.

1 The Model

In this chapter we are concerned with modeling population growth. Typically we start the population with a single individual. This individual has a random number of offspring. Each child has itself a number of offspring and so on. The first question is about survival. Can such a process survive forever? Can we compute the survival probability? In order to be able to do computations we need to make some assumptions. We will assume that different individuals have independent but identical offspring distributions. Hence, all individuals are put in the same condition (offspring wise) and we get a branching (or cascading) effect. We will concentrate on biological applications. This process turns out to be also a good model for a number of physical phenomena.

This stochastic process was introduced independently by Bienaymé (who got the mathematics right and was forgotten for many years) and by Galton and Watson (who got the mathematics wrong but got their names attached to this process) to model the survival of family names. An initial set of individuals which we call the zeroth generation have a number of offspring that are called the first generation; their offspring are called the second generation and so on. We denote the size of the nth generation by Z_n, $n \geq 0$.

We now give the mathematical definition of the Bienaymé–Galton–Watson (BGW) process $(Z_n)_{n \geq 0}$. The state space S of $(Z_n)_{n \geq 0}$ is the set of positive (including zero) integers. We suppose that each individual gives birth to Y particles in the next generation where Y is a positive integer-valued random variable with distribution $(p_k)_{k \geq 0}$. In other words

© Springer Science+Business Media New York 2014
R.B. Schinazi, *Classical and Spatial Stochastic Processes: With Applications to Biology*, DOI 10.1007/978-1-4939-1869-0_2

$$P(Y = k) = p_k, \text{ for } k = 0, 1, \dots.$$

Moreover we assume that the number of offspring of the various individuals in the various generations are chosen independently according to the distribution $(p_k)_{k \geq 0}$.

The process is governed by the so-called one-step transition probabilities

$$p(i, j) = P(Z_{n+1} = j | Z_n = i).$$

That is, $p(i, j)$ is the conditional probability that $Z_{n+1} = j$ given that $Z_n = i$. We also have

$$p(0, i) = 0 \text{ if } i \geq 1 \text{ and } p(0, 0) = 1.$$

That is, once the process is at 0 (or extinct) it stays there. State 0 (no individuals) is also said to be an absorbing state (or trap) for $(Z_n)_{n \geq 0}$.

Observe that

$$p(i, j) = P(Z_{n+1} = j | Z_n = i) = P\left(\sum_{k=1}^{i} Y_k = j\right) \text{ for } i \geq 1, j \geq 0,$$

where $(Y_k)_{1 \leq k \leq i}$ is a sequence of independent identically distributed (i.i.d.) random variables with distribution $(p_k)_{k \geq 0}$. This shows that the distribution of Z_{n+1} can be computed using the distribution of Z_n only. That is, there is no need to know the complete history of the process given by $Z_0, Z_1, \dots, Z_n$ in order to compute the distribution of Z_{n+1}. It is enough to know Z_n. This is called the Markov property.

A word on notation. For $Z_n = i$ we should have written Z_{n+1} as

$$Z_{n+1} = \sum_{k=1}^{i} Y_{k,n} = j$$

where $(Y_{k,n})_{1 \leq k \leq i}$ is an i.i.d. sequence. The subscript n is to indicate that we use a different independent sequence for every n. We omit the n in the notation to avoid a double index.

Let the mean offspring be

$$m = \sum_{k=0}^{\infty} k p_k,$$

where m is possibly $+\infty$ if the series does not converge. Let q be the probability that the BGW process starting from a single individual eventually dies out. We also introduce the generating function of the offspring distribution

$$f(s) = \sum_{k=0}^{\infty} p_k s^k \text{ for } |s| \leq 1.$$

We now state the main result of this chapter. The process will be said to survive if there exists at least one individual for every generation n. Mathematically, surviving means

$$\{Z_n \geq 1, \text{ for all } n \geq 0\}.$$

Theorem 1.1. *Let $(Z_n)_{n\geq0}$ be a BGW process with offspring distribution $(p_k)_{k\geq0}$. Assume that $p_0 + p_1 < 1$.*
If $m \leq 1$, then $P(Z_n \geq 1, \text{ for all } n \geq 0|Z_0 = 1) = 0$.
If $m > 1$, there exists q in $[0, 1)$ such that $P(Z_n \geq 1, \text{ for all } n \geq 0|Z_0 = 1) = 1 - q > 0$. Moreover q, the extinction probability, is the unique solution in $[0, 1)$ of the equation $f(s) = s$ when $m > 1$.

The process BGW is said to be subcritical, critical, and supercritical according to whether $m < 1$, $m = 1$, or $m > 1$. Observe that the BGW process may survive forever if and only if $m > 1$. So the only relevant parameter of the offspring distribution for survival is m. However, the probability $1 - q$ of surviving forever depends on the whole distribution $(p_k)_{k\geq1}$ through its generating function.

The proof of Theorem 1.1 will be given in the last section of this chapter.

It is useful to have a graphical representation of the process $(Z_n)_{n\geq0}$. See Fig. 2.1. Survival of the process corresponds to an infinite tree. Death of a process corresponds to a finite tree.

We now apply Theorem 1.1 to a few examples.

Example 1.1. Consider a BGW process with the offspring distribution $P(Y = 0) = p_0 = 1/6$, $P(Y = 1) = p_1 = 1/2$, and $P(Y = 2) = p_2 = 1/3$. We first compute the average offspring per individual.

$$m = E(Y) = 7/6 > 1.$$

So the survival probability $1 - q$ is strictly positive in this case. The generating function of Y is

$$f(s) = 1/6 + s/2 + s^2/3.$$

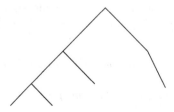

Fig. 2.1 This is a graphical representation of a BGW. The process starts with one individual at the top of the tree. We see that $Z_0 = 1$, $Z_1 = 2$, $Z_2 = 3$ and $Z_3 = 4$

The extinction probability q is the only solution strictly less than 1 of $f(s) = s$. This equation can be written as

$$\frac{1}{6} - \frac{1}{2}s + \frac{1}{3}s^2 = \frac{1}{3}(s-1)(s-\frac{1}{2}) = 0.$$

There are two solutions $s = 1$ and $s = 1/2$. We know that q is the solution in $[0, 1)$. Hence, $q = 1/2$. So starting with a single particle there is a probability $1/2$ that the process will survive forever.

Example 1.2. Lotka (1939) has used a geometric distribution to fit the offspring of the American male population. He found that

$$p_0 = P(Y = 0) = 1/2 \text{ and } p_i = P(Y = i) = (\frac{3}{5})^{i-1}\frac{1}{5} \text{ for } i \geq 1,$$

where Y represents the number of sons that a male has in his lifetime. Recall that

$$\sum_{n\geq 0} x^n = \frac{1}{1-x} \qquad \sum_{n\geq 1} nx^{n-1} = \frac{1}{(1-x)^2} \text{ for } |x| < 1.$$

So

$$m = \sum_{n\geq 1} np_n = \sum_{n\geq 1} n(\frac{3}{5})^{n-1}\frac{1}{5} = \frac{5}{4} > 1.$$

Hence, the extinction probability is strictly less than 1 and is a solution of $f(s) = s$ where

$$f(s) = \frac{1}{2} + \sum_{n\geq 1}(\frac{3}{5})^{n-1}\frac{1}{5}s^n = \frac{1}{2} + \frac{s}{5-3s}.$$

Solving the equation $f(s) = s$ yields

$$\frac{3}{5}s^2 - \frac{11}{10}s + \frac{1}{2} = 0.$$

As always $s = 1$ is a solution. The unique root strictly less than 1 is $q = 5/6$. So under this model a given male has a probability of $1/6$ of generating a family that survives forever.

Proposition 1.1. *Assume that m (the mean offspring) is finite. We have*

$$E(Z_n|Z_0 = 1) = m^n \text{ for } n \geq 0.$$

Proof of Proposition 1.1. We do a proof by induction. Note that

$$E(Z_1|Z_0 = 1) = E(Y) = m = m^1.$$

Hence, the formula holds for $n = 1$. Assume now that it holds for n. By conditioning on Z_n we get

$$E(Z_{n+1}|Z_0 = 1) = \sum_{k \geq 1} E(Z_{n+1}|Z_n = k)P(Z_n = k|Z_0 = 1),$$

where by the Markov property we are using that

$$E(Z_{n+1}|Z_0 = 1, Z_n = k) = E(Z_{n+1}|Z_n = k).$$

For every $k \geq 1$,

$$E(Z_{n+1}|Z_n = k) = E\left(\sum_{i=1}^{k} Y_i\right) = km.$$

Thus,

$$E(Z_{n+1}|Z_0 = 1) = \sum_{k \geq 1} km P(Z_n = k|Z_0 = 1) = mE(Z_n|Z_0 = 1).$$

Since by the induction hypothesis we have $E(Z_n|Z_0 = 1) = m^n$ we can conclude that

$$E(Z_{n+1}|Z_0 = 1) = mm^n = m^{n+1}.$$

This completes the proof of Proposition 1.1.

Problems

1. Consider a BGW process with offspring distribution

$$P(Y = 0) = 1 - p \text{ and } P(Y = 2) = p$$

where p is a fixed parameter in $(0, 1)$.

(a) Compute the mean offspring m.
(b) For which p does the process have a positive probability of surviving?
(c) Sketch the extinction probability q as a function of p.

2. Consider a BGW with offspring distribution $(p_k)_{k \geq 0}$ and extinction probability q.

(a) Show that $q \geq p_0$.
(b) Show that $q = 0$ (survival is certain) if and only if $p_0 = 0$.

3. Let r be in $(0, 1)$. Consider a BGW with the offspring distribution

$$p_k = (1 - r)r^k \text{ for } k \geq 0.$$

Find the extinction probability q as a function of r.

4. Find the extinction probability q if the offspring distribution $(p_k)_{k \geq 0}$ is given by

$$p_k = \binom{3}{k} (1/2)^3$$

for $k = 0, 1, 2, 3$.

5. Consider a BGW with mean offspring m.

(a) Show that

$$P(Z_n \geq 1) \leq E(Z_n).$$

(b) Assume that the mean offspring $m = 1/2$. Show that the probability that Z_n has survived ten generations is less than $\frac{1}{2^{10}}$.

6. Redo Example 1.2 with a truncated geometric distribution. More precisely, take $p_0 = P(Y = 0) = 1/2$ and

$$p_i = P(Y = i) = (\frac{3}{5})^{i-1}c \text{ for } 1 \leq i \leq 10.$$

(a) Find c.
(b) Compute (approximately) the extinction probability.

7. Find approximately the extinction probability in the case $p_k = e^{-2}2^k/k!$ for $k \geq 0$.

8. Discuss the behavior of Z_n in the case $p_0 + p_1 = 1$.

9. Consider an isolated island where the original stock of surnames is 100. Assume that each surname has an extinction probability $q = 9/10$.

(a) After many generations how many surnames do you expect in the island?
(b) Do you expect the total population of the island to be increasing or decreasing?

10. Consider a supercritical BGW for which the extinction probability $q = 1/2$. Start the process with five particles. What is the probability that the process will survive forever?

11. Consider a supercritical BGW with offspring distribution $(p_k)_{k \geq 0}$. Let the extinction probability be q.

(a) Show that for every integer $k \geq 0$ we have

$$P(\text{ extinction}|Z_1 = k) = q^k.$$

(b) Show that

$$P(Z_1 = k|Z_0 = 1) = p_k.$$

(c) Use (a) and (b) to show that

$$P(\text{ extinction}|Z_0 = 1) = \sum_{k=0}^{\infty} p_k q^k.$$

(d) Use (c) to show that q is a solution of the equation $f(s) = s$ where f is the generating function of the offspring distribution.

12. Let f be the generating function of the probability distribution $(p_k)_{k \geq 0}$.

(a) Show that f is increasing on $[0, 1]$.
(b) Show that $f(0) \geq 0$ and $f(1) = 1$.
(c) Show that f is concave up on $[0, 1]$.
(d) Graph three functions with properties (a), (b), and (c) and $f'(1) < 1$, $f'(1) = 1$ and $f'(1) > 1$, respectively.
(e) Using the graphs in (d) show that the equation $f(s) = s$ has a solution $0 \leq q < 1$ if and only if $f'(1) > 1$.
(f) How does (e) relate to Theorem 1.1?

13. Consider a BGW process Z_n with offspring distribution

$$P(Y = 0) = 1 - p \text{ and } P(Y = 2) = p$$

where p is a fixed parameter in $(0, 1)$. A simulation of the offspring distribution Y with $p = 3/4$ has yielded the following observations 2,2,0,2,2,0,2,2,2,2,0.

(a) Use the simulation to graph the corresponding BGW tree.
(b) Use the simulation to find Z_n for as many n as possible.

14. Consider the following algorithm. Let U be a continuous uniform random variable on $(0, 1)$. If $U < p$ set $Y = 2$, if $Y > p$ set $Y = 0$.
  Show that this algorithm generates a random variable Y with the distribution

$$P(Y = 0) = 1 - p, P(Y = 2) = p.$$

(Recall that $P(U < x) = x$ for x in $(0, 1)$.)

15. Use Problem 13 to simulate a BGW process Z_n with offspring distribution

$$P(Y = 0) = 1 - p \text{ and } P(Y = 2) = p,$$

for $p = 1/4$, $p = 1/2$ and $p = 3/4$. Do multiple runs. Interpret the results.

2   The Probability of a Mutation in a Branching Process

Consider the following question. We have a population modeled by a BGW process. Each time an individual is born in the population it has a probability μ of having a certain mutation. We want to compute the probability that this mutation will eventually appear in the population. We will actually do the computation for a particular BGW and we will need several steps.

2.1   An Equation for the Total Progeny Distribution

We will start by finding the probability distribution of

$$X = \sum_{n \geq 0} Z_n = 1 + \sum_{n \geq 1} Z_n.$$

The random variable X counts all the births that ever occur in the population (i.e., the total progeny) as well as the founding individual (as usual we are taking $Z_0 = 1$). Our first goal is to find the distribution of X. The main step is to find the generating function g of X. Let

$$g(s) = \sum_{k \geq 1} s^k P(X = k) = E(s^X).$$

The sum above starts at 1 since X is always at least 1. We know that a generating function is always defined (i.e., the power series converges) for s in $[0, 1]$.

We will find an equation for g by conditioning on the first generation Z_1. We have

$$g(s) = E(s^X) = \sum_{k \geq 0} E(s^X | Z_1 = k) P(Z_1 = k | Z_0 = 1) \tag{2.1}$$

Given that $Z_1 = k$ we have k individuals at time 1 starting k independent BGW. Hence, X is the sum of the founding individual and all the individuals in these k BGW. Moreover, for each one of these k BGW the total number of births (plus the founding individual) has the same distribution as X. Therefore,

$$E(s^X | Z_1 = k) = E(s^{1 + X_1 + X_2 + \cdots + X_k})$$

where the X_i, $1 \leq i \leq k$, are independent random variables with the same distribution as X. Thus, by independence

$$E(s^X | Z_1 = k) = s E(s^{X_1}) E(s^{X_2}) \ldots E(s^{X_k}).$$

Since the X_i, $1 \leq i \leq k$, have the same distribution as X we get

$$E(s^X | Z_1 = k) = sE(s^X)^k.$$

Using the last equality in (2.1) yields

$$g(s) = E(s^X) = \sum_{k \geq 0} sE(s^X)^k P(Z_1 = k | Z_0 = 1) = s \sum_{k \geq 0} g(s)^k p_k$$

since

$$P(Z_1 = k | Z_0 = 1) = P(Y = k) = p_k.$$

Hence,

$$g(s) = s \sum_{k \geq 0} p_k g(s)^k.$$

Observe now that the generating function f of the offspring distribution is

$$f(s) = \sum_{k \geq 0} p_k s^k.$$

Therefore,

$$g(s) = sf(g(s)) \tag{2.2}$$

This is an elegant equation for g but for most offspring distributions (represented by f) it cannot be solved. Next we solve the equation in a particular case.

2.2 The Total Progeny Distribution in a Particular Case

Consider the following offspring distribution: $p_0 = 1 - p$ and $p_2 = p$ where p is a fixed parameter in $[0, 1]$. That is, for $n \geq 0$, every individual of the nth generation gives birth to two individuals with probability p or does not give birth at all with probability $1 - p$. This yields the $(n + 1)$th generation. Biologically, we are thinking of a population of bacteria or virus that either divide into two individuals (corresponding to births in the model) or die.

Recall that g is the generating function of X:

$$g(s) = \sum_{n=1}^{\infty} P(X = 2n - 1 | Z_0 = 1)s^{2n-1}.$$

We take into account only the odd values for X because the births that occur always occur by pairs and since we start with a single individual the random variable X is odd.

By (2.2) g is the solution of the equation

$$g(s) = sf(g(s)),$$

where f is the generating function of the offspring distribution. For this particular case

$$f(s) = 1 - p + ps^2.$$

Hence,

$$g(s) = s(1 - p + pg(s)^2),$$

and

$$psg(s)^2 - g(s) + s(1 - p) = 0.$$

For a fixed s let $g(s) = u$, we get

$$psu^2 - u + s(1 - p) = 0$$

and

$$u^2 - \frac{1}{ps}u + \frac{s(1 - p)}{ps} = 0.$$

This is a quadratic equation in u. We complete the square to get

$$(u - \frac{1}{2ps})^2 - \frac{1}{4p^2s^2} + \frac{s(1 - p)}{ps} = 0.$$

Hence,

$$(u - \frac{1}{2ps})^2 = \frac{1}{4p^2s^2} - \frac{s(1 - p)}{ps} = \frac{1 - 4ps^2(1 - p)}{4p^2s^2}.$$

Note that $1 - 4ps^2(1 - p) > 0$ if and only if

$$s^2 < \frac{1}{4p(1 - p)}.$$

Since $4p(1 - p) \leq 1$ for p in $(0, 1)$ (why?) we have that $\frac{1}{4p(1-p)} \geq 1$. Hence, for s in $(0, 1)$ we have

$$s^2 < \frac{1}{4p(1 - p)}.$$

Therefore, we have two solutions for $u = g(s)$

$$u_1 = \frac{1}{2ps} - \sqrt{\frac{1 - 4ps^2(1 - p)}{4p^2s^2}} \text{ or } u_2 = \frac{1}{2ps} + \sqrt{\frac{1 - 4ps^2(1 - p)}{4p^2s^2}}.$$

We need to decide which expression is $g(s)$. Observe that

$$u_2 = \frac{1}{2ps} + \sqrt{\frac{1 - 4ps^2(1 - p)}{4p^2s^2}} \sim \frac{1}{4ps}$$

where $f \sim g$ means that

$$\lim_{s \to 0} \frac{f(s)}{g(s)} = 1.$$

Since $\frac{1}{4ps}$ is not bounded near 0 and g is defined at 0 (in fact $g(0) = 0$) $g(s)$ cannot be u_2. It must be u_1. Hence,

$$g(s) = \frac{1}{2ps} - \sqrt{\frac{1 - 4ps^2(1 - p)}{4p^2s^2}} \tag{2.3}$$

Since

$$g(s) = \sum_{n=1}^{\infty} P(X = 2n - 1 | Z_0 = 1)s^{2n-1}$$

the expression in (2.3) is useful in computing the distribution of X only if we can find a power series expansion for that expression. This is our next task.

Using the binomial expansion from Calculus we have the following lemma.

Lemma 2.1. *For $|x| < 1$ we have*

$$1 - \sqrt{1 - x} = \sum_{n=1}^{\infty} c_n x^n$$

where

$$c_n = \frac{(2n - 2)!}{2^{2n-1}n!(n - 1)!} \text{ for } n \geq 1.$$

Lemma 2.1 will be proved in the exercises. Going back to (2.3) we have

$$g(s) = \frac{1}{2ps} - \sqrt{\frac{1 - 4ps^2(1 - p)}{4p^2s^2}} = \frac{1}{2ps}(1 - \sqrt{1 - 4ps^2(1 - p)}).$$

Let $x = 4ps^2(1 - p)$ in Lemma 2.1 to get

$$g(s) = \frac{1}{2ps} \sum_{n=1}^{\infty} c_n (4p(1 - p)s^2)^n = \sum_{n=1}^{\infty} \frac{(2n - 2)!}{n!(n - 1)!}(1 - p)^n p^{n-1} s^{2n-1}.$$

This power series expansion shows that g has no singularity at $s = 0$. In fact, it is infinitely differentiable at any s in $(-1, 1)$. We now have a power series expansion for g and this yields the distribution of X. We get for $n \geq 1$ that

$$P(X = 2n - 1 | Z_0 = 1) = \frac{(2n - 2)!}{n!(n - 1)!}(1 - p)^n p^{n-1}.$$

2.3   The Probability of a Mutation

We are finally ready to compute the probability that a certain mutation eventually appears in a population modeled by a BGW. We will do the computation for the particular BGW we have been considering so far. Recall that in this particular case the offspring distribution is $p_0 = 1 - p$ and $p_2 = p$. We also assume that at each birth in the BGW there is a probability μ that the new individual has the mutation. Moreover, the new individual has the mutation (or not) independently of everything else. Let M be the event that the mutation eventually appears in the population. Let M^c be the complement of M, that is, the event that the mutation never appears in the population. Let

$$X = \sum_{n \geq 0} Z_n = 1 + \sum_{n \geq 1} Z_n,$$

where the BGW starts with one individual, that is, $Z_0 = 1$. Note that X can be infinite. In fact, the process $(Z_n)_{n \geq 0}$ survives if and only if $X = +\infty$ (why?). Intuitively it is clear that if there are infinitely many births and each birth has a fixed probability μ of a mutation independently of all other births then the mutation will occur with probability 1. We will examine this issue more carefully in the last subsection of this section.

The probability that the mutation never appears in the population is therefore

$$P(M^c) = P(M^c; X < +\infty).$$

By the rule of averages we have

$$P(M^c) = \sum_{k=1}^{\infty} P(M^c | X = k) P(X = k) = \sum_{k=1}^{\infty} (1 - \mu)^{k-1} P(X = k).$$

Recall the total progeny distribution

$$P(X = 2n - 1 | Z_0 = 1) = \frac{(2n - 2)!}{n!(n - 1)!}(1 - p)^n p^{n-1}.$$

Hence,

$$P(M^c) = \sum_{n=1}^{\infty} (1 - \mu)^{2n-2} P(X = 2n - 1 | Z_0 = 1)$$

$$= \sum_{n=1}^{\infty} (1 - \mu)^{2n-2} \frac{(2n - 2)!}{n!(n - 1)!}(1 - p)^n p^{n-1}.$$

To sum this series we will use Lemma 2.1. We first rearrange the general term of the series to get

$$P(M^c) = \frac{1}{2p(1 - \mu)^2} \sum_{n=1}^{\infty} \frac{(2n - 2)!}{2^{2n-1}n!(n - 1)!}\left(4(1 - \mu)^2(1 - p)p\right)^n.$$

Therefore,

$$P(M^c) = \frac{1}{2p(1 - \mu)^2} \sum_{n=1}^{\infty} c_n \left(4(1 - \mu)^2(1 - p)p\right)^n$$

where the sequence c_n is defined in Lemma 2.1. By that lemma we have

$$P(M^c) = \frac{1}{2p(1 - \mu)^2}\left(1 - \sqrt{1 - 4p(1 - p)(1 - \mu)^2}\right) \tag{2.4}$$

This is the probability no mutation ever appears in the population. Of course, the probability that a mutation does eventually appear is $P(M) = 1 - P(M^c)$.

2.4 Application: The Probability of Drug Resistance

Drug resistance is a constant threat to the health of individuals who are being treated for a variety of ailments: HIV, tuberculosis, cancer to cite a few. It is also a threat to the population as a whole since there is a risk that a treatable disease may be replaced by a non-treatable one. This is the case, for instance, for tuberculosis.

We are interested in evaluating the risk of a treatment induced drug resistance. In the presence of a drug the drug sensitive strain is weakened (how much it is weakened depends on the efficacy of the drug) and this gives an edge to the drug resistant strain if it appears before the drug is able to eradicate all pathogens. Therefore, what determines the treatment outcome is whether total eradication takes place before the appearance of a drug resistance mutation. We propose a model to compute the probability of pathogen eradication before drug resistance appears.

We now recall the model we have been studying in the previous subsections. We assume that at every unit time a given pathogen may die with probability $1 - p$ or divide in two with probability p. Thus, the mean offspring per pathogen is $2p$. We assume that p is strictly between 0 and 1. If $2p > 1$, then there is a positive probability for the family tree of a single drug sensitive pathogen to survive forever. If $2p \leq 1$, then eradication is certain for drug sensitive pathogens. The parameter p is a measure of efficacy of the drug. The smaller the p the more efficient the drug is and the more likely eradication of the drug sensitive pathogen is.

As always for branching processes, we assume that the number of pathogens each pathogen gives birth to is independent of the number of pathogens any other pathogen gives birth to at the same time. We also assume that for each birth of pathogen there is a probability μ that the new pathogen is drug resistant. We denote by N the number of pathogens at the beginning of treatment.

Recall from (2.4) above that the probability of no mutation starting with a single pathogen is

$$P(M^c|Z_0 = 1) = \frac{1}{2p(1-\mu)^2}\left(1 - \sqrt{1 - 4p(1-p)(1-\mu)^2}\right) \qquad (2.5)$$

Usually the treatment will start when the patient has some symptoms. These symptoms start when the number of pathogens is high enough. Therefore we are interested in the model for $Z_0 = N$ where N is a rather large number. Define the function f as

$$f(N, \mu, p) = P(M^c|Z_0 = N).$$

That is, f is the probability that the drug resistance mutation never appears given that the treatment starts with N pathogens. In our model each of the N pathogens starts its own independent BGW. Hence, the probability that there is no mutation in the population is the probability that none of the N independent BGW generate a mutation. Therefore,

$$f(N, \mu, p) = f(1, \mu, p)^N = \left(\frac{1}{2p(1-\mu)^2}\left(1 - \sqrt{1 - 4p(1-p)(1-\mu)^2}\right)\right)^N.$$

We are now going to see that the function f behavior changes drastically depending whether $p < 1/2$ or $p > 1/2$.

- Subcritical case: $p < 1/2$. In order to obtain a friendlier expression for f we compute a linear approximation in μ as μ approaches 0. Note that the linear approximation for $(1 - \mu)^{-2}$ is $1 + 2\mu$. A little algebra shows

$$\sqrt{1 - 4p(1 - p)(1 - \mu)^2} = |1 - 2p|\sqrt{1 + \frac{8p(1 - p)}{(1 - 2p)^2}\mu - \frac{4p(1 - p)}{(1 - 2p)^2}\mu^2}.$$

By the binomial expansion

$$\sqrt{1 + x} \sim 1 + \frac{1}{2}x$$

where $f \sim g$ means that

$$\lim_{x \to 0} \frac{f(x)}{g(x)} = 1.$$

Hence,

$$\sqrt{1 - 4p(1 - p)(1 - \mu)^2} \sim |1 - 2p|(1 + \frac{4p(1 - p)}{(1 - 2p)^2}\mu)$$

as $\mu \to 0$. Thus, for $p < 1/2$ we have the linear approximation

$$f(1, \mu, p) \sim 1 - \frac{2p}{1 - 2p}\mu.$$

Since $f(N, \mu, p) = f(1, \mu, p)^N$ we have for $p < 1/2$

$$f(N, \mu, p) \sim (1 - \frac{2p}{1 - 2p}\mu)^N \sim \exp(-\frac{2p}{1 - 2p}N\mu) \tag{2.6}$$

where we are using that

$$(1 - x)^N \sim \exp(-Nx)$$

as x approaches 0. Formula (2.6) tells us that the critical parameter for a successful drug treatment is $N\mu$. The smaller $N\mu$ the larger $f(N, \mu, p)$ and therefore the larger the probability of no drug resistance. The model confirms what has been found by experience. For HIV for instance better start the treatment early (smaller N) than late (larger N). It also has been found that it is better to use simultaneously three drugs rather than one. The probability μ of the appearance of a mutation which is resistant to all three drugs is much smaller than the probability of the appearance of a mutation which is resistant to a single drug. The model also suggests that it is not necessary to have both N and μ small. It is enough to have $N\mu$ small.

- Supercritical case: $p > 1/2$. In this case too we may do an approximation similar to what was done in the subcritical case. But more importantly we have for $p > 1/2$

$$f(N, \mu, p) \leq (\frac{1-p}{p})^N \qquad (2.7)$$

We will prove (2.7) below. First note that $\frac{1-p}{p} < 1$ (for $p > 1/2$) and N is very large. Therefore, $f(N, \mu, p)$ will be very small. That is, drug resistance will almost certainly appear. In this case, the model suggests that treatment is futile at best. The drug will make appear something worse (a drug resistant strain) than what it is supposed to cure.

We now prove (2.7). Starting with one pathogen the probability that the BGW will go extinct is $\frac{1-p}{p}$. See the exercises. Starting with N pathogens in order for the drug resistant mutation not to appear it is necessary (but not sufficient) that all N independent BGW to go extinct. For if one the BGW survives forever the mutation is certain to appear (in a BGW that survives forever there are infinitely many births and each one has the constant probability $\mu > 0$ of being resistant). Hence, the event "Drug resistance does not occur" is included in "All N BGW go extinct." Therefore, the probability of the first event (i.e. $f(N, \mu, p)$) is less than the probability of the second event. Now the probability that N independent BGW go extinct is $(\frac{1-p}{p})^N$. This proves (2.7).

2.5 Application: Cancer Risk

Cancer has long been thought to appear after several successive mutations. We assume here that a cancerous cell appears after two successive mutations. We consider a tissue in the human body (a tissue is an ensemble of similar cells that together carry out a specific function). The cells of this tissue undergo a fixed number D of divisions over the lifetime of the tissue. We also assume that there is a probability μ_1 per division of producing a cell with a type 1 mutation. A cell carrying a type 1 mutation is a pre-cancerous cell. If a type 1 cell appears, it starts a BGW process. At each unit time each cell in this BGW may die with probability $1 - p_1$ or divide in two type 1 cells with probability p_1. At each division of a type 1 cell there is a probability μ_2 for each daughter cell that it be a type 2 cell. A type 2 cell is a cancerous cell. We are interested in computing the probability that a cancerous cell appear over the lifetime of the tissue.

In order for a cancerous cell to appear we first need a type 1 mutation and then a type 2 mutation appearing in the BGW started by the type 1 cell. Assume that at each of the D divisions we have the same probability p that the two successive mutations appear. Assuming also independence of these D events we get

$$P(\text{no cancer}) = (1 - p)^D.$$

We now compute p. At a given cell division let A_1 be the event of a first mutation and let A_2 be the event that a second mutation eventually occurs. The probability of $A_1 \cap A_2$ is exactly p. Therefore,

$$p = P(A_1 \cap A_2) = P(A_1)P(A_2|A_1).$$

We know that $P(A_1) = \mu_1$ and $P(A_2|A_1)$ is the probability that a mutation occurs in a BGW starting with a single individual with mutation probability μ_2 and division probability p_1. Hence,

$$P(A_2|A_1) = 1 - f(1, \mu_2, p_1)$$

where

$$f(1, \mu_2, p_1) = \frac{1}{2p_1(1 - \mu_2)^2}\left(1 - \sqrt{1 - 4p_1(1 - p_1)(1 - \mu_2)^2}\right),$$

has been computed in Sect. 2.4. Recall that $f(1, \mu, p)$ is the probability that no mutation occurs in a BGW with mutation probability μ and division probability p. So $1 - f(1, \mu_2, p_1)$ is the probability that a mutation does occur. Therefore,

$$p = \mu_1(1 - f(1, \mu_2, p_1)).$$

Let

$$S(p_1, \mu_2) = 1 - f(1, \mu_2, p_1).$$

Hence,

$$P(\text{no cancer}) = (1 - p)^D \sim \exp(-pD) = \exp(-\ell S(p_1, \mu_2))$$

where $\ell = \mu_1 D$ and the approximation holds for p approaching 0. The formula above is interesting in several ways. It shows that μ_1 and D are important only through their product ℓ. Moreover, the parameter ℓ determines whether p_1 and μ_2 are important. We now see why.

- Small ℓ. Note that $S(p_1, \mu_2)$ is a probability and is therefore in $[0, 1]$ for all p_1 and μ_2. Hence, $\ell S(p_1, \mu_2) \leq \ell$ and

$$P(\text{no cancer}) \geq \exp(-\ell) \sim 1 - \ell$$

where the approximation holds for ℓ approaching 0. That is, the risk of cancer is of order ℓ. The parameters p_1 and μ_2 are almost irrelevant. This is so because if ℓ is small then the first mutation is unlikely during the lifetime of the tissue. If the first mutation is unlikely, then so is the second mutation since the second mutation can only happen on top of the first.

- Large ℓ. If ℓ is large, then it is quite likely that a first mutation will occur during the lifetime of the tissue. Whether the second mutation occurs depends on p_1 and μ_2. It turns out that what determines whether the second mutation occurs is p_1. The parameter μ_2 is not really relevant. See the problems.

2.6 The Total Progeny May Be Infinite

We now revisit the distribution of the total progeny computed in Sect. 2.2. Recall that the total progeny X is defined by

$$X = \sum_{n \geq 0} Z_n = 1 + \sum_{n \geq 1} Z_n,$$

where the BGW $(Z_n)_{n \geq 0}$ starts with one individual, that is, $Z_0 = 1$. If the distribution of X is correct, then $\sum_{n=1}^{\infty} P(X = 2n - 1|Z_0 = 1)$ should be 1. Or should it? We now do the computation.

$$\sum_{n=1}^{\infty} P(X = 2n - 1|Z_0 = 1) = \sum_{n=1}^{\infty} \frac{(2n-2)!}{n!(n-1)!}(1-p)^n p^{n-1}.$$

We will use Lemma 2.1 to compute this infinite series. In order to do so we rearrange the general term of the series. We have

$$\frac{(2n-2)!}{n!(n-1)!}(1-p)^n p^{n-1} = \frac{(2n-2)!}{2^{2n-1}n!(n-1)!}2^{2n-1}(1-p)^n p^n p^{-1}$$

$$= \frac{(2n-2)!}{2^{2n-1}n!(n-1)!}(4p(1-p))^n (2p)^{-1}.$$

Hence,

$$\sum_{n=1}^{\infty} P(X = 2n - 1|Z_0 = 1) = \frac{1}{2p} \sum_{n=1}^{\infty} \frac{(2n-2)!}{2^{2n-1}n!(n-1)!}(4p(1-p))^n$$

$$= \frac{1}{2p} \sum_{n=1}^{\infty} c_n (4p(1-p))^n$$

where the sequence $(c_n)_{n \geq 1}$ is defined in Lemma 2.1. We now let $x = 4p(1-p)$ in Lemma 2.1 to get

$$\sum_{n=1}^{\infty} P(X = 2n - 1|Z_0 = 1) = \frac{1}{2p}(1 - \sqrt{1 - 4p(1-p)}).$$

Note that $1 - 4p(1 - p) = (1 - 2p)^2$. Hence,

$$\sqrt{1 - 4p(1 - p)} = |1 - 2p|.$$

Therefore,

$$\sum_{n=1}^{\infty} P(X = 2n - 1 | Z_0 = 1) = \frac{1}{2p}(1 - |1 - 2p|).$$

There are two cases to consider. If $p \leq 1/2$ then $|1 - 2p| = 1 - 2p$ and

$$\sum_{n=1}^{\infty} P(X = 2n - 1 | Z_0 = 1) = 1$$

as expected. However, if $p > 1/2$, then $|1 - 2p| = -1 + 2p$ and

$$\sum_{n=1}^{\infty} P(X = 2n - 1 | Z_0 = 1) = \frac{1}{2p}(2 - 2p) = \frac{1 - p}{p}$$

which is not 1 (except when $p = 1/2$)! What is going on? Recall our definition of X.

$$X = \sum_{n \geq 0} Z_n$$

where $(Z_n)_{n \geq 0}$ is a BGW. Now note that the Z_n are positive or 0 integers. So X is finite if and only if $Z_n = 0$ for all n larger than some fixed integer (why?). This is the same as saying that X is finite if and only if the BGW $(Z_n)_{n \geq 0}$ dies out. It is easy to check (see the exercises) that for this particular BGW extinction occurs if and only if $p \leq 1/2$. If $p > 1/2$, then there is a positive probability that the BGW does not die out. That is, there is a positive probability that X is infinite. Observe also that $\sum_{n=1}^{\infty} P(X = 2n - 1 | Z_0 = 1)$ is the probability that X takes a finite value. This series does not include the possibility that X is infinite. This is why when $p > 1/2$ the series is strictly less than 1. We have

$$\sum_{n=1}^{\infty} P(X = 2n - 1 | Z_0 = 1) = P(X < +\infty | Z_0 = 1) = \frac{1 - p}{p}.$$

We will check in the exercises that if $p > 1/2$ the probability that $(Z_n)_{n \geq 0}$ dies out is indeed $\frac{1-p}{p}$.

We now tie another loose end. What is the probability that a mutation appears when X is infinite? We will show below that this probability is 1. Let M be the probability that a mutation occurs at some point in the process $(Z_n)_{n \geq 0}$. Let k be a positive integer. We have

$$P(M^c; X \geq k) = P(M^c | X \geq k) P(X \geq k) \leq (1 - \mu)^{k-1} P(X \geq k) \qquad (2.8)$$

This is so because given $X \geq k$ there are at least $k - 1$ births in the population and each birth has independently failed to carry the mutation. Observe also that the sequence $\{X \geq k\}$ for $k \geq 1$ is decreasing. That is, for $k \geq 1$

$$\{X \geq k + 1\} \subset \{X \geq k\}.$$

Hence, by Proposition 1.1 in the appendix

$$\lim_{k \to \infty} P(X \geq k) = P(\bigcap_{k \geq 1} \{X \geq k\}).$$

Note that if $X \geq k$ for every $k \geq 1$ then $X = +\infty$. Thus,

$$\lim_{k \to \infty} P(X \geq k) = P(X = +\infty).$$

With a similar argument we show that

$$\lim_{k \to \infty} P(M^c; X \geq k) = P(M^c; X = +\infty).$$

Letting k go to infinity in (2.8) yields

$$P(M^c; X = +\infty) \leq P(X = +\infty) \lim_{k \to \infty} (1 - \mu)^{k-1} = 0$$

since $0 < 1 - \mu < 1$. Hence, a mutation occurs with probability 1.

Problems

1. Let

$$X = \sum_{n \geq 0} Z_n = 1 + \sum_{n \geq 1} Z_n,$$

where the BGW $(Z_n)_{n \geq 0}$ starts with one individual, that is, $Z_0 = 1$. Show that the process $(Z_n)_{n \geq 0}$ survives if and only if $X = +\infty$.

2. Consider a BGW with the following offspring distribution: $p_0 = 1 - p$ and $p_2 = p$ where p is a fixed parameter in $[0, 1]$.

(a) Show that the BGW may survive if and only $p > 1/2$.
(b) Show that the moment generating function of the offspring distribution is

$$f(s) = 1 - p + ps^2.$$

(c) Show that the extinction probability is $\frac{1-p}{p}$ when $p > 1/2$.

3. Consider a BGW with the offspring distribution

$$p_k = (1 - r)r^k \text{ for } k \geq 0,$$

where r is in $(0, 1)$.

(a) Show that the generating function of the offspring distribution is for $s \leq 1$

$$f(s) = \frac{1 - r}{1 - rs}.$$

(b) Let g be the generating function of the total progeny for this BGW starting with a single individual. Use Eq. (2.2) to show that

$$rg(s)^2 - g(s) + s(1 - r) = 0.$$

(c) The quadratic equation in (b) has two solutions. Explain why $g(s)$ is in fact

$$g(s) = \frac{1}{2r}(1 - \sqrt{1 - 4s(1 - r)r}).$$

(d) Use Lemma 2.1 to show that

$$g(s) = \frac{1}{2r} \sum_{n \geq 1} c_n 4^n (1 - r)^n r^n s^n$$

where c_n is defined in Lemma 2.1.

(e) Let X be the total progeny of this BGW starting with a single individual. Show that

$$P(X = n | Z_0 = 1) = \frac{1}{2r} c_n 4^n (1 - r)^n r^n.$$

(f) Use (e) and Lemma 2.1 to show that

$$\sum_{n \geq 1} P(X = n | Z_0 = 1) = \frac{1}{2r}(1 - |1 - 2r|).$$

(g) Show that $\sum_{n \geq 1} P(X = n) = 1$ for $r \leq 1/2$ and $\sum_{n \geq 1} P(X = n) < 1$ for $r > 1/2$. Could you have known that without computing the distribution of X?

(h) Set $r = 1/4$. Compute $P(X = n)$ for $n = 1, 2 \ldots 10$.

(i) Set $r = 3/4$. Compute $P(X = n)$ for $n = 1, 2 \ldots 10$.

4. In this problem we compute the probability of a given mutation for the BGW studied in Problem 3. Let M^c be the event that no mutation ever appears in the BGW that started with a single individual. Assume that for each birth in the BGW there is a probability μ that the new individual has the mutation.

(a) Show that

$$P(M^c|Z_0 = 1) = \sum_{n \geq 1} P(X = n|Z_0 = 1)(1 - \mu)^{n-1}$$

where X is the total progeny of the BGW.

(b) Use Problem 2 (e) and Lemma 2.1 in (a) to show that

$$P(M^c|Z_0 = 1) = \frac{1}{2r(1 - \mu)}(1 - \sqrt{1 - 4(1 - r)r(1 - \mu)}).$$

(c) Explain why for every integer $N \geq 1$

$$P(M^c|Z_0 = N) = P(M^c|Z_0 = 1)^N.$$

(d) Set $r = 1/10$. Compute $P(M^c|Z_0 = N)$ for several values of N and μ for which $N\mu = 1$. What do these computations suggest?

(e) Set $r = 6/10$ and $N = 10$. Compute $P(M^c|Z_0 = N)$ for $\mu = 10^{-4}$, $\mu = 10^{-5}$, $\mu = 10^{-6}$. What do these computations suggest?

5. Recall from Calculus the binomial expansion. Let α be a real number. Then, for all x in $(-1, 1)$,

$$(1 + x)^\alpha = 1 + \sum_{k=1}^{\infty} a_k x^k$$

where

$$a_k = \frac{\alpha(\alpha - 1) \ldots (\alpha - k + 1)}{k!}$$

for $k \geq 1$. Use the binomial expansion (in the case $\alpha = 1/2$) to prove Lemma 2.1. Do a proof by induction.

6. For $p < 1/2$ we have approximated

$$f(N, \mu, p) = f(1, \mu, p)^N = \left(\frac{1}{2p(1 - \mu)^2} \left(1 - \sqrt{1 - 4p(1 - p)(1 - \mu)^2}\right) \right)^N$$

by using

$$h(N, \mu, p) = \exp(-\frac{2p}{1 - 2p} N\mu)$$

as $\mu \to 0$.

How good is the approximation? Compute f and h for p in $[0.1, 0.4]$, μ in $[10^{-8}, 10^{-4}]$ and N in $[10, 10^6]$. Find out the maximal error for this approximation.

7. The cells of a tissue undergo a fixed number D of divisions over the lifetime of the tissue. Assume that there is a probability μ_1 per division of producing a cell with a type 1 mutation.

(a) Show that the probability of having at least one type 1 mutation over the lifetime of the tissue is

$$s = 1 - (1 - \mu_1)^D.$$

(b) Show that

$$s \sim 1 - \exp(-\mu_1 D)$$

as μ_1 approaches 0.

(c) Let $m = \mu_1 D$. Sketch the graph of s as a function of m.

8. In the cancer risk model of Sect. 2.6 we have shown that the risk r of cancer for a certain tissue is

$$r \sim 1 - \exp(-\ell S(p_1, \mu_2)).$$

Let $\mu_2 = 10^{-6}$. Sketch the graphs of r as a function of p_1 for $\ell = 0.01$, $\ell = 0.1$, and $\ell = 1$. Interpret these graphs.

3 Proof of Theorem 1.1

This proof involves mostly analysis arguments and is not important for the sequel. We include it for the sake of completeness.

Before proving Theorem 1.1 we will need a few properties of generating functions. Recall that the generating function of the probability distribution $(p_k)_{k \geq 0}$ is

$$f(s) = \sum_{k \geq 0} p_k s^k.$$

We have seen already that a generating function is defined on $(-1, 1)$. Since f is also defined at 1 and we are only interested in positive numbers we will take the domain of f to be $[0, 1]$.

An useful Analysis lemma is the following.

Lemma 3.1. *Let* $(b_n)_{n \geq 0}$ *be a positive sequence and let*

$$g(t) = \sum_{n \geq 0} b_n t^n.$$

Assume that g is defined on $[0, 1)$. *Then*

$$\lim_{t \to 1^-} g(t) = \sum_{n \geq 0} b_n$$

where both sides are possibly $+\infty$.

For a proof see Proposition A1.9 in the Appendix of "Theoretical Probability for applications" by S.C. Port.

Applying Lemma 3.1 to the generating function f we see that

$$\lim_{s \to 1^-} f(s) = \sum_{n \geq 0} p_n = 1.$$

Since $f(1) = 1$, f is left continuous at 1. On the other hand, a power series is infinitely differentiable (and hence continuous) on any open interval where it is defined. Therefore f is continuous on $[0, 1]$.

We will need another application of Lemma 3.1. As noted above the function f is differentiable on $(0, 1)$ and since a power series can be differentiated term by term

$$f'(s) = \sum_{n \geq 1} n p_n s^{n-1}.$$

By Lemma 3.1

$$\lim_{s \to 1^-} f'(s) = \sum_{n \geq 1} n p_n = m$$

where $\lim_{s \to 1^-} f'(s)$ and m may be both infinite.

We now go back to the BGW process and compute the generating function of Z_n for $n \geq 1$.

Proposition 3.1. *Let* $f_1 = f$ *and* $f_{n+1} = f \circ f_n$ *for* $n \geq 1$. *For* $n \geq 1$, *the generating function of* Z_n *conditioned on* $Z_0 = 1$ *is* f_n.

Proof of Proposition 3.1. We prove this by induction. Let g_n be the generating function of Z_n given that $Z_0 = 1$. We have

$$g_1(s) = E(s^{Z_1} | Z_0 = 1) = E(s^Y) = f(s) = f_1(s),$$

so the property holds for $n = 1$. Assume that $g_n = f_n$. Given $Z_n = k$, the distribution of Z_{n+1} is the same as the distribution of $\sum_{i=1}^{k} Y_i$ where the Y_i are i.i.d. with distribution $(p_k)_{k \geq 0}$. Hence,

$$E(s^{Z_{n+1}} | Z_n = k) = E(s^{\sum_{i=1}^{k} Y_i}) = E(s^{Y_1}) E(s^{Y_2}) \dots E(s^{Y_k}) = (E(s^{Y_1}))^k = f(s)^k.$$

By the Markov property

$$g_{n+1}(s) = E(s^{Z_{n+1}}|Z_0 = 1)$$

$$= \sum_{k=0}^{\infty} E(s^{Z_{n+1}}|Z_n = k)P(Z_n = k|Z_0 = 1)$$

$$= \sum_{k=0}^{\infty} P(Z_n = k|Z_0 = 1)f(s)^k$$

$$= g_n(f(s))$$

and by our induction hypothesis we get $g_{n+1} = g_n \circ f = f_n \circ f = f_{n+1}$. This completes the proof of Proposition 3.1.

We now prove Theorem 1.1. We start by dealing with the easiest case: $m < 1$. For any positive integer valued random variable X

$$E(X) = \sum_{k \geq 0} kP(X = k) \geq \sum_{k \geq 1} P(X = k) = P(X \geq 1).$$

Hence,

$$P(X \geq 1) \leq E(X).$$

We use the preceding inequality and Proposition 1.1 to get

$$P(Z_n \geq 1|Z_0 = 1) \leq E(Z_n|Z_0 = 1) = m^n.$$

Since $m < 1$

$$\lim_{n \to \infty} P(Z_n \geq 1|Z_0 = 1) = 0,$$

and the convergence occurs exponentially fast. Observe that since 0 is a trap for $(Z_n)_{n \geq 0}$ the sequence of events $\{Z_n \geq 1\}$ is decreasing. That is,

$$\{Z_{n+1} \geq 1\} \subset \{Z_n \geq 1\}.$$

In words, if $Z_{n+1} \geq 1$ then we must have $Z_n \geq 1$ (why?).

By Proposition 1.1 in the Appendix

$$\lim_{n \to \infty} P(Z_n \geq 1|Z_0 = 1) = P(\bigcap_{n \geq 0}\{Z_n \geq 1\}|Z_0 = 1) =$$

$$P(Z_n \geq 1 \text{ for all } n \geq 0|Z_0 = 1) = 0.$$

This proves Theorem 1.1 in the case $m < 1$.

For the cases $m = 1$ and $m > 1$ we will need the following observations. For any positive integer valued random variable X we can define the generating function g_X on $[0, 1]$ by

$$g_X(s) = \sum_{k=0}^{\infty} P(X = k)s^k.$$

If we let $s = 0$, we get $g_X(0) = P(X = 0)$.

Since f_n is the moment generating function of Z_n conditioned on $\{Z_0 = 1\}$ we get

$$P(Z_n = 0|Z_0 = 1) = f_n(0),$$

and since the sequence of events $\{Z_n = 0\}$ is increasing (why?) we have by Proposition 1.1 in the Appendix

$$\lim_{n \to \infty} f_n(0) = P(\bigcup_{n \geq 1} \{Z_n = 0\}|Z_0 = 1) \qquad (3.1)$$

Let q to be the probability of extinction. Define

$$q = P(Z_n = 0 \text{ for some } n \geq 1|Z_0 = 1).$$

Observe that

$$\{Z_n = 0 \text{ for some } n \geq 1\} = \bigcup_{n \geq 1} \{Z_n = 0\}.$$

Hence, by (3.1)

$$q = P(Z_n = 0 \text{ for some } n \geq 1|Z_0 = 1) = P(\bigcup_{n \geq 1} \{Z_n = 0\}|Z_0 = 1) = \lim_{n \to \infty} f_n(0).$$

Therefore,

$$q = \lim_{n \to \infty} f_n(0) \qquad (3.2)$$

Now

$$f_{n+1}(0) = f(f_n(0)) \qquad (3.3)$$

Since $f_{n+1}(0)$ and $f_n(0)$ both converge to q and f is continuous on $[0, 1]$ we get $f(q) = q$ by letting n go to infinity in (3.3). That is, q is a fixed point of f.

Our task now will be to show that depending on m we will have $q = 1$ (extinction is certain) or $q < 1$ (survival has positive probability).

We first consider $m = 1$.

$$f'(s) = \sum_{k \geq 1} k p_k s^{k-1} < \sum_{k \geq 1} k p_k = m = 1 \text{ for } s < 1.$$

Therefore, for any $s < 1$, by the Mean Value Theorem there is a c in $(s, 1)$ such that

$$f(1) - f(s) = f'(c)(1 - s) < 1 - s,$$

and so for any $s < 1$

$$f(s) > s.$$

Therefore there is no solution to the equation $f(s) = s$ in the interval $[0,1]$ other than $s = 1$. Hence, we must have $q = 1$. This completes the proof of Theorem 1.1 for the case $m = 1$.

Consider now $m > 1$. We have for s in $[0, 1)$ that

$$f'(s) = \sum_{k=1}^{\infty} k p_k s^{k-1}.$$

Moreover, by Lemma 3.1 we have that

$$\lim_{s \to 1} f'(s) = \sum_{k=1}^{\infty} k p_k = m.$$

Hence, there exists an $\eta > 0$ such that if $s \geq 1 - \eta$ then $1 < f'(s) < m$. By the Mean Value Theorem, for any s in $(1 - \eta, 1)$ there is a c in $(s, 1)$ such that

$$f(1) - f(s) = (1 - s) f'(c).$$

Since $f(1) = 1$ and $f'(c) > 1$ (since $c > s > 1 - \eta$) we have $1 - f(s) > 1 - s$. Hence, there is an $\eta > 0$ such that

$$f(s) < s \text{ for } s \text{ in } (1 - \eta, 1) \tag{3.4}$$

Let $g(x) = x - f(x)$. This is a continuous function on $[0, 1]$. By (3.4) $g(s) > 0$ for $s > 1 - \eta$. Moreover, $f(0) \geq 0$ and therefore $g(0) \leq 0$. By the Intermediate Value Theorem we have at least one solution in $[0, 1 - \eta) \subset [0, 1)$ to the equation $g(s) = 0$ or equivalently to the equation $f(s) = s$. Denote this solution by s_1.

We now show that there is a unique solution to the equation $f(s) = s$ in $[0,1)$. By contradiction assume there is another solution t_1 in $[0,1)$. Assume without loss of generality that $s_1 < t_1$. Since $f(1) = 1$ we have at least three solutions to the equation $g(s) = 0$ on $[0, 1]$. We apply Rolle's Theorem on $[s_1, t_1]$ and on $[t_1, 1]$ to get the existence of ξ_1 in (s_1, t_1) and ξ_2 in $(t_1, 1)$ such that $g'(\xi_1) = g'(\xi_2) = 0$. Hence, $f'(\xi_1) = f'(\xi_2) = 1$. Observe that

$$f''(s) = \sum_{k \geq 2} k(k-1) p_k s^{k-2}.$$

Since $p_0 + p_1 < 1$ we must have $p_k > 0$ for at least one $k \geq 2$ (why?). Therefore $f''(s) > 0$ for s in $(0,1)$ and f' is strictly increasing on $(0, 1)$. Therefore, we cannot have $\xi_1 < \xi_2$ and $f'(\xi_1) = f'(\xi_2) = 1$. We have reached a contradiction. Hence, there is a unique solution to the equation $f(s) = s$ in $[0,1)$.

At this point we know that $q = s_1$ or $q = 1$ since these are the two only solutions of $f(s) = s$ in $[0,1]$. By contradiction assume that $q = 1$. By (3.2), $\lim_{n \to \infty} f_n(0) = q = 1$. Hence, for n large enough, $f_n(0) > 1 - \eta$. By (3.4) (let $s = f_n(0)$ there) this implies that $f(f_n(0)) < f_n(0)$. That is, $f_{n+1}(0) < f_n(0)$. But this contradicts the fact that $(f_n(0))_{n \geq 1}$ is an increasing sequence. Hence q cannot be 1. It must be the unique solution of $f(s) = s$ which is strictly less than 1. This completes the proof of Theorem 1.1.

Problems

1. Show that for every $n \geq 1$ we have

$$\{Z_{n+1} \geq 1\} \subset \{Z_n \geq 1\}.$$

2. Show that for every $n \geq 1$ we have

$$\{Z_n = 0\} \subset \{Z_{n+1} = 0\}.$$

3. Consider a probability distribution $(p_k)_{k \geq 0}$. Show that if $p_0 + p_1 < 1$ we must have $p_k > 0$ for at least one $k \geq 2$.

4. Let f be the generating function of the probability distribution $(p_k)_{k \geq 0}$.

(a) Show that if $p_0 < 1$ then f is strictly increasing on $(0, 1)$.
(b) Show that if $p_0 + p_1 < 1$ then f' is strictly increasing on $(0, 1)$.

5. Consider the generating function f of the offspring distribution $(p_k)_{k \geq 0}$. We assume that $p_0 + p_1 < 1$ and therefore f is strictly increasing. Assume also that the mean offspring distribution $m > 1$.

(a) We have shown in (3.4) that $f(s) < s$ for all s in $(1 - \eta, 1)$ where η is some positive real number. Show that in fact

$$f(s) < s \text{ for all } s \text{ in } (q, 1).$$

(Do a proof by contradiction. Show that if $f(s_0) \geq s_0$ for some s_0 in $(q, 1)$ then the equation $f(s) = s$ would have at least two solutions in $[0, 1)$.)

(b) Recall that f_n is the nth iterate of the generating function f (see Proposition 3.1). For a fixed s in $(q, 1)$ define the sequence $a_n = f_n(s)$ for $n \geq 1$. Show that for every $n \geq 1$

$$a_n > q.$$

(c) Show that the sequence a_n is decreasing.

(d) Show that a_n converges to a limit ℓ which is in $[q, 1)$. Show also that $f(\ell) = \ell$.

(e) Show that ℓ is in fact q. That is, we have shown that for any s in $(q, 1)$, $f_n(s)$ converges to q.

(f) Do steps similar to (a) through (e) to show that $f_n(s)$ converges to q for any s in $[0, q)$.

6. By Proposition 3.1 f_n is the generating function of Z_n. That is, for $n \geq 1$

$$f_n(s) = \sum_{k \geq 0} P(Z_n = k | Z_0 = 1)s^k.$$

(a) In Problem 5 we proved that $f_n(s)$ converges to q for any fixed s in $[0, 1)$. Show that

$$\lim_{n \to \infty} P(Z_n = 0 | Z_0 = 1) = q.$$

(b) Show that for any fixed $k \geq 1$

$$\lim_{n \to \infty} P(Z_n = k | Z_0 = 1) = 0.$$

(c) If Z_n does not get extinct, where does it go to?

Notes

For a history of the Bienaymé–Galton–Watson process see Kendall (1975). There are entire books on branching processes. My favorite is Harris (1989). The drug resistance application is based on Schinazi (2006a,b), see also Iwasa et al. (2004) for another stochastic model. See Nowak and May (2000) for differential equation models for drug resistance. The cancer risk application is based on Schinazi (2006a,b). Durrett et al. (2009) and Schweinsberg (2008) have also studied stochastic models for multi-stage carcinogenesis but their models lead to rather involved mathematics. Schweinsberg (2008) gives a nice account of the history of the multi-stage carcinogenesis hypothesis.

References

Durrett, R., Schmidt, D., Schweinsberg, J.: A waiting time problem arising from the study of multistage carcinogenesis. Ann. Appl. Probab. **19**, 676–718 (2009)

Harris, T.E.: The Theory of Branching Processes. Dover, New York (1989)

Iwasa, Y., Michor, F., Nowak, M.A.: Evolutionary dynamics of invasion and escape. J. Theor. Biol. **226**, 205–214 (2004)

Kendall, D.G.: The genealogy of genealogy branching processes before (and after) 1873. Bull. Lond. Math. Soc. **7**(3), 225–253 (1975)

Lotka, A.J.: Theorie analytique des associations biologiques. Actualites scientifiques et industrielles, Hermann, Paris **780** (1939)

Nowak, M.A., May, R.M.: Virus Dynamics. Oxford University Press, New York (2000)

Port, S.C.: Theoretical Probability for Applications. Wiley, New York (1994)

Schinazi, R.B.: The probability of treatment induced drug resistance. Acta Biotheor. **54**, 13–19 (2006a)

Schinazi, R.B.: A stochastic model for cancer risk. Genetics **174**, 545–547 (2006b)

Schweinsberg, J.: Waiting for m mutations. Electron. J. Probab. **13**, 1442–1478 (2008)

Chapter 3
The Simple Symmetric Random Walk

A simple random walk is a discrete time stochastic process $(S_n)_{n \geq 0}$ on the integers $\mathbf{Z}$. Let

$$S_0 = 0$$

and for $n \geq 1$

$$S_n = S_{n-1} + X_n,$$

where $X_1, X_2, \ldots$ is a sequence of independent identically distributed random variables with the following distribution,

$$P(X = 1) = p \text{ and } P(X = -1) = q = 1 - p.$$

Hence, $S_n = S_{n-1} + 1$ with probability p and $S_n = S_{n-1} - 1$ with probability q. In this chapter we consider the symmetric case $p = q = 1/2$.

Here are two possible interpretations. We may think of S_n as the winnings (possibly negative) of a certain gambler after n one dollar bets. For each bet the gambler wins \$1 with probability p and loses \$1 with probability q. Another (more poetic) interpretation is to think of S_n as the position of a walker after n steps on $\mathbf{Z}$. The walker takes one step to the right with probability p or one step to the left with probability q.

Given the simplicity of the model random walks have a surprisingly rich mathematical behavior and can be used to model many different questions.

© Springer Science+Business Media New York 2014
R.B. Schinazi, *Classical and Spatial Stochastic Processes: With Applications to Biology*, DOI 10.1007/978-1-4939-1869-0_3

1   Graphical Representation

Consider the simple symmetric random walk $(S_n)_{n \geq 0}$ on $\mathbf{Z}$. The event $\{S_n = s\}$ is represented in the plane by the point (n, s). Starting with $S_0 = 0$ (which is represented by the point $(0, 0)$) there are a number of ways to get $S_n = s$. We now give an example.

Example 1.1. What outcomes yield $S_4 = 2$?

First note that the gambler must have won 3 bets and must have lost 1 bet. There are four ways to achieve that:

$$(S_0, S_1, S_2, S_3, S_4) = (0, -1, 0, 1, 2)$$
$$(S_0, S_1, S_2, S_3, S_4) = (0, 1, 0, 1, 2)$$
$$(S_0, S_1, S_2, S_3, S_4) = (0, 1, 2, 1, 2)$$
$$(S_0, S_1, S_2, S_3, S_4) = (0, 1, 2, 3, 2)$$

Each one of these four ways corresponds to a certain path in the plane. For instance, $(S_0, S_1, S_2, S_3, S_4) = (0, 1, 0, 1, 2)$ corresponds to the path in Fig. 3.1.

We now give a formal definition of a path.

Definition 1.1. Let $n \leq m$ be two positive integers. Let s and t be two integers. A *path* from (n, s) to (m, t) is a sequence of points $(i, s_i)_{n \leq i \leq m}$ such that $s_n = s$, $s_m = t$, and $|s_i - s_{i+1}| = 1$ for all integers i in $[n, m-1]$.

Let n and k be two integers such that $0 \leq k \leq n$. Recall the definition of the binomial coefficient

$$\binom{n}{k} = \frac{n!}{k!(n-k)!}$$

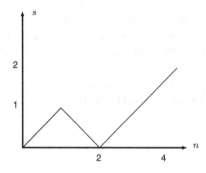

Fig. 3.1 This is one of the four possible paths from $(0, 0)$ to $(4, 2)$

where if $k \geq 1$

$$k! = 1 \times 2 \times \ldots \times k$$

and 0! equals 1.

Recall also that the number of ways to place k identical objects in n different spots is given by $\binom{n}{k}$. Going back to Example 1.1, for $S_4 = 2$ we have three upwards steps and one downwards step. The number of ways to position the three upwards steps among a total of four steps is

$$\binom{4}{3} = 4$$

If we know in which position the upwards steps are we also know in which position the downwards step is. This determines a unique path. Hence, there are four possible paths from $(0,0)$ to $(4,2)$.

We are now ready to count the number of paths from the origin to some fixed point.

Proposition 1.1. *Let s and $n \geq 0$ be integers. If $n + s$ is even, then the number of paths from $(0,0)$ to (n,s) is*

$$\binom{n}{\frac{n+s}{2}}$$

If $n + s$ is odd, then there are no paths going from $(0,0)$ to (n,s). For instance, there are no paths going from $(0,0)$ to $(3,0)$ (why?).

Observe also that in order to go from $(0,0)$ to (n,s) we must have $-n \leq s \leq n$.

Proof of Proposition 1.1. Any path from $(0,0)$ to (n,s) has exactly n steps. Each step in the path is either $+1$ or -1. Let $n^+ \geq 0$ be the number of $+1$ steps and $n^- \geq 0$ be the number of -1 steps. Since the total number of steps is n we must have

$$n = n^+ + n^-.$$

After n steps the path has reached $n^+ - n^-$ (why?). Hence,

$$s = n^+ - n^-.$$

Solving for n^+ and n^- we get

$$n^+ = \frac{n+s}{2} \text{ and } n^- = \frac{n-s}{2}.$$

The solutions n^+ and n^- need to be positive integers. Hence, $n + s$ must be even.

As noted above a path is uniquely determined by the position of its $+1$ (or upwards) steps. Hence, there are

$$\binom{n}{n^+}$$

paths going from $(0, 0)$ to (n, s). We could have used the downwards steps instead of the upwards steps. It is easy to check that

$$\binom{n}{n^+} = \binom{n}{n^-},$$

see the problems. This completes the proof of Proposition 1.1.

Consider a path with points

$$(n, s_1), (n + 1, s_2) \ldots (m, s_{m-n}).$$

This path is said to *intersect* the n-axis if and only if there is at least one i in $[1, m - n]$ such that $s_i = 0$.

A very useful tool to count the number of paths in certain situations is the following so-called *Reflection Principle*.

Proposition 1.2. *Let $0 \leq n < m$ and $0 \leq s < t$ be four integers. Let A, A', and C have coordinates (n, s), $(n, -s)$, and (m, t), respectively. The number of paths from A to C that intersect the n axis equals the total number of paths from A' to C (see Fig. 3.2).*

Proof of Proposition 1.2. Consider a path from A to C that intersects the n axis. It is a sequence of points with coordinates $(n, s_1), (n + 1, s_2) \ldots (m, s_{m-n})$ such that $s_1 = s$ and $s_{m-n} = t$ and $|s_i - s_{i+1}| = 1$ for all integers i in $[1, m - n]$. Moreover, since the path intersects the n axis we must have $s_i = 0$ for at least one i. Let b be the smallest of such indices. That is,

$$b = \min\{i \in [1, m - n] : s_i = 0\}.$$

Let the point B have coordinates $(b, 0)$, see Fig. 3.2. Consider now the path

$$(n, -s_1), (n + 1, -s_2) \ldots (b, 0)(b + 1, s_{b+1}) \ldots (m, s_{m-n}).$$

That is, we take the reflection of the first b points (with respect to the n-axis) and keep the other points as they are to get a new path going from A' to C.

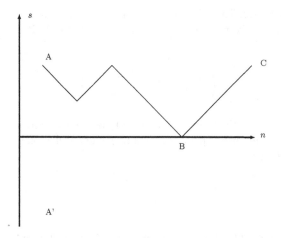

Fig. 3.2 A has coordinates (n, s), A' has coordinates $(n, -s)$, and B has coordinates (m, t). A' is the reflection of A with respect to the n axis. There are as many paths from A to C that intersect the n axis as the total number of paths from A' to C

This construction shows that for every path from A to C that intersects the n axis corresponds a path from A' to C. Conversely we can start with a path from A' to C. It must intersect the n axis and we may define the point B as above. By taking the reflection of the points up to B we get a new path going from A to C. Hence, there is one-to-one correspondence between the paths from A to C that intersect the n axis and the paths from A' to C. This proves Proposition 1.2.

We now give a first application of the Reflection Principle.

Proposition 1.3. *Let* $0 < s < n$ *such that* $n + s$ *is even. The number of paths from* $(0, 0)$ *to* (n, s) *such that* $s_0 = 0, s_1 > 0, s_2 > 0 \ldots, s_n = s$ *is*

$$\binom{n-1}{\frac{n+s}{2} - 1} - \binom{n-1}{\frac{n+s}{2}} = \frac{s}{n}\binom{n}{\frac{n+s}{2}}$$

Paths with the property described in Proposition 1.3 will be said to be in the *positive quadrant.*

Note that if $s = n$ there is only one path from $(0, 0)$ to (n, s) since all steps need to be upwards.

By Proposition 1.1. the total number of paths from $(0, 0)$ to (n, s) is $\binom{n}{\frac{n+s}{2}}$. Hence, by Proposition 1.3 the fraction of paths that stay in the positive quadrant is $\frac{s}{n}$.

Proof of Proposition 1.3. We are counting the number of paths that stay in the positive quadrant. The first step must be upwards. Therefore the second point in the path must be $(1, 1)$. The total number of paths from $(1, 1)$ to (n, s) equals the total number of paths from $(0, 0)$ to $(n - 1, s - 1)$ (why?). By Proposition 1.1 this number is

$$\binom{n-1}{\frac{n-1+s-1}{2}} = \binom{n-1}{\frac{n+s}{2} - 1}$$

But we want only the paths from $(1, 1)$ to (n, s) that do not intersect the n axis. By Proposition 1.2 the number of paths we want to exclude is the total number of paths from $(1, -1)$ to (n, s). That is,

$$\binom{n-1}{\frac{n+s}{2}}$$

Therefore, the number of paths that stay in the positive quadrant is

$$\binom{n-1}{\frac{n+s}{2} - 1} - \binom{n-1}{\frac{n+s}{2}}$$

It is now easy to show that this difference can also be written as $\frac{s}{n} \binom{n}{\frac{n+s}{2}}$. See the problems. This concludes the proof of Proposition 1.3.

Example 1.2. Going back to Example 1.1, how many paths go from $(0, 0)$ to $(4, 2)$ and stay in the positive quadrant?

As we already saw the total number of paths from $(0, 0)$ to $(4, 2)$ is 4. The fraction that stay in the positive quadrant is $\frac{s}{n} = \frac{2}{4}$. Hence, there are two such paths. We can check this result by direct inspection. The two paths are

$$(S_0, S_1, S_2, S_3, S_4) = (0, 1, 2, 1, 2)$$
$$(S_0, S_1, S_2, S_3, S_4) = (0, 1, 2, 3, 2)$$

2   Returns to 0

So far we have counted paths. Now we are going to use the path counting to compute probabilities. Consider all the paths starting at $(0, 0)$ with n steps. At each step there are two choices (upwards or downwards), hence there are 2^n such paths. Since upwards and downwards steps are equally likely all 2^n paths are equally likely. That is, the probability of any given path is $\frac{1}{2^n}$.

Example 2.1. As we have seen in Example 1.1 there are four paths from $(0,0)$ to $(4,2)$. The paths from $(0,0)$ to $(4,2)$ form a subset of all possible paths with four steps starting at $(0,0)$. Hence,

$$P(S_4 = 2|S_0 = 0) = \frac{4}{2^4} = \frac{1}{4}.$$

On the other hand if we want the probability of reaching $(2,4)$ only through paths in the positive quadrant we have only two such paths. Hence,

$$P(S_1 > 0, S_2 > 0, S_3 > 0, S_4 = 2|S_0 = 0) = \frac{2}{2^4} = \frac{1}{8}.$$

We are particularly interested in paths that return to 0. Returns to 0 can only happen at even times. Let

$$u_{2n} = P(S_{2n} = 0|S_0 = 0) = P_0(S_{2n} = 0).$$

By Proposition 1.1 there are $\binom{2n}{n}$ paths from $(0,0)$ to $(2n,0)$. Each path has $2n$ steps. Hence,

$$u_{2n} = \frac{1}{2^{2n}} \binom{2n}{n}$$

for all $n \geq 0$. Note that $u_0 = 1$.

As the next three propositions will show probabilities of more involved events can be computed using u_{2n}. These propositions will be helpful in the finer analysis of random walks that will be performed in the next section. Their proofs illustrate the power of the graphical method and of the reflection principle.

Proposition 2.1. *We have for all $n \geq 1$*

$$P(S_1 > 0, S_2 > 0, \ldots, S_{2n} > 0|S_0 = 0) = \frac{1}{2}u_{2n}.$$

Proof of Proposition 2.1. Observe that S_{2n} is even and can be anywhere from 2 to $2n$ if $S_{2n} > 0$. Hence,

$$P_0(S_1 > 0, S_2 > 0, \ldots, S_{2n} > 0) = \sum_{k=1}^{n} P_0(S_1 > 0, S_2 > 0, \ldots, S_{2n} = 2k),$$

where P_0 denotes the conditioning on $S_0 = 0$. If $k \leq n - 1$, we can use Proposition 1.3 (where $2n$ plays the role of n and $2k$ plays the role of s) to get that

$$P_0(S_1 > 0, S_2 > 0, \ldots, S_{2n} = 2k) = \frac{1}{2^{2n}}\left(\binom{2n-1}{n+k-1} - \binom{2n-1}{n+k}\right)$$

For $1 \le k \le n$ let

$$a_k = \frac{1}{2^{2n-1}}\binom{2n-1}{n+k-1}$$

Note that a_k depends on k and n. A better (but more cumbersome) notation is $a_{k,n}$. We choose to omit n in the notation since n remains constant in our computation below. For $k \le n-1$

$$P_0(S_1 > 0, S_2 > 0, \ldots, S_{2n} = 2k) = \frac{1}{2}\left(\frac{1}{2^{2n-1}}\binom{2n-1}{n+k-1} - \frac{1}{2^{2n-1}}\binom{2n-1}{n+k}\right)$$

$$= \frac{1}{2}(a_k - a_{k+1}).$$

On the other hand, if $k = n$, then

$$P_0(S_1 > 0, S_2 > 0, \ldots, S_{2n} = 2n) = \frac{1}{2^{2n}}.$$

Note that the r.h.s. equals $\frac{1}{2}a_n$. Therefore,

$$P_0(S_1 > 0, S_2 > 0, \ldots, S_{2n} > 0) = \sum_{k=1}^{n} P_0(S_1 > 0, S_2 > 0, \ldots, S_{2n} = 2k)$$

$$= \frac{1}{2}\sum_{k=1}^{n-1}(a_k - a_{k+1}) + \frac{1}{2}a_n.$$

All the terms in the previous sum cancel except for the first one. Thus,

$$P_0(S_1 > 0, S_2 > 0, \ldots, S_{2n} > 0) = \frac{1}{2}a_1.$$

A little algebra shows that $a_1 = u_{2n}$. This completes the proof of Proposition 2.1.

Proposition 2.2. *We have for all $n \ge 1$*

$$P_0(S_1 \ne 0, S_2 \ne 0, \ldots, S_{2n} \ne 0) = u_{2n}$$

Proof of Proposition 2.2. Observe that

$$P_0(S_1 \ne 0, S_2 \ne 0, \ldots, S_{2n} \ne 0) = P_0(S_1 > 0, S_2 > 0, \ldots, S_{2n} > 0)$$
$$+ P_0(S_1 < 0, S_2 < 0, \ldots, S_{2n} < 0).$$

Note now that there as many paths corresponding to the event $\{S_0 = 0, S_1 > 0, S_2 > 0, \ldots, S_{2n} > 0\}$ as there are corresponding to $\{S_0 = 0, S_1 < 0, S_2 < 0, \ldots, S_{2n} < 0\}$ (why?). Hence, by Proposition 2.1

$$P_0(S_1 \neq 0, S_2 \neq 0, \ldots, S_{2n} \neq 0) = 2P_0(S_1 > 0, S_2 > 0, \ldots, S_{2n} > 0) = u_{2n}.$$

This completes the proof of Proposition 2.2.

Proposition 2.3. *We have for all $n \geq 1$*

$$P_0(S_1 \geq 0, S_2 \geq 0, \ldots, S_{2n} \geq 0) = u_{2n}$$

Proof of Proposition 2.3. This proof is similar to the proof of Proposition 2.1. Note first that

$$P_0(S_1 \geq 0, S_2 \geq 0, \ldots, S_{2n} \geq 0) = P_0(S_1 > -1, S_2 > -1, \ldots, S_{2n} > -1).$$

We have

$$P_0(S_1 > -1, S_2 > -1, \ldots, S_{2n} > -1) = \sum_{k=0}^{n} P_0(S_1 > -1, S_2 > -1, \ldots, S_{2n} = 2k).$$

The event $\{S_0 = 0, S_1 > -1, S_2 > -1, \ldots, S_{2n} = 2k\}$ corresponds to paths from $(0,0)$ to $(2n, 2k)$ that do not intersect the $s = -1$ axis. We can easily adapt the Reflection Principle to this situation. Doing a reflection with respect to the line $s = -1$ shows that the number of paths from $(0,0)$ to $(2n, 2k)$ that do intersect the $s = -1$ axis equals the total number of paths from $(0, -2)$ to $(2n, 2k)$. This in turn is the same as the total number of paths from $(0,0)$ to $(2n, 2k + 2)$. Therefore, the number of paths from $(0,0)$ to $(2n, 2k)$ that do not intersect the $s = -1$ axis is

$$\binom{2n}{n+k} - \binom{2n}{n+k+1}$$

for $k \leq n - 1$. For $k = n$ there is only one such path. Hence,

$$P_0(S_1 > -1, S_2 > -1, \ldots, S_{2n} > -1) = \sum_{k=0}^{n} P_0(S_1 > -1, S_2 > -1, \ldots, S_{2n} = 2k)$$

$$= \sum_{k=0}^{n-1} \frac{1}{2^{2n}} \left(\binom{2n}{n+k} - \binom{2n}{n+k+1} \right) + \frac{1}{2^{2n}}$$

It is easy to see that all the terms in the last sum cancel except for the first one. That is,

$$P_0(S_1 > -1, S_2 > -1, \ldots, S_{2n} > -1) = \frac{1}{2^{2n}} \binom{2n}{n} = u_{2n}.$$

This completes the proof of Proposition 2.3.

3   Recurrence

We now turn to the probability of *first* return to the origin. As noted before a return to the origin can only occur at even times. The probability f_{2n} of a first return to the origin at time $2n$ is defined for $n = 1$ by

$$f_2 = P(S_2 = 0 | S_0 = 0)$$

and for $n \geq 2$ by

$$f_{2n} = P(S_2 \neq 0, S_4 \neq 0, \ldots, S_{2n-2} \neq 0, S_{2n} = 0 | S_0 = 0).$$

It turns out that f_{2n} can be expressed using u_{2n}. Recall that u_{2n} is the probability of a return (not necessarily the first!) to 0 at time $2n$. We know that

$$u_{2n} = P(S_{2n} = 0 | S_0 = 0) = \frac{1}{2^{2n}} \binom{2n}{n}$$

and that $u_0 = 1$.

Proposition 3.1. *We have for all $n \geq 1$*

$$f_{2n} = u_{2n-2} - u_{2n}.$$

Proof of Proposition 3.1. We first recall an elementary probability fact. We denote the complement of B (everything not in B) by B^c.

Lemma 3.1. *Assume that A and B are two events such that $B \subset A$. Then*

$$P(A \cap B^c) = P(A) - P(B).$$

Proof of Lemma 3.1. We have

$$P(A) = P(A \cap B) + P(A \cap B^c).$$

Since $B \subset A$ then $A \cap B = B$. Hence,

$$P(A) = P(B) + P(A \cap B^c).$$

That is,

$$P(A \cap B^c) = P(A) - P(B).$$

This completes the proof of Lemma 3.1.

Back to the proof of Proposition 3.1. Consider first the case $n \geq 2$. Define the events A and B by

$$A = \{S_2 \neq 0, S_4 \neq 0, \ldots, S_{2n-2} \neq 0\}$$
$$B = \{S_2 \neq 0, S_4 \neq 0, \ldots, S_{2n-2} \neq 0, S_{2n} \neq 0\}.$$

Note that

$$A \cap B^c = \{S_2 \neq 0, S_4 \neq 0, \ldots, S_{2n-2} \neq 0, S_{2n} = 0\}.$$

Since $B \subset A$ Lemma 3.1 applies and we have

$$P(A \cap B^c) = P(A) - P(B).$$

Note that

$$P(A \cap B^c) = f_{2n}.$$

By Proposition 2.2

$$P(A) = u_{2n-2}$$

and

$$P(B) = u_{2n}.$$

Hence, for all $n \geq 2$

$$f_{2n} = u_{2n-2} - u_{2n}.$$

We now deal with $n = 1$. Note that

$$f_2 = u_2 = P(S_2 = 0 | S_0 = 0).$$

Moreover, it is easy to see that $P(S_2 = 0 | S_0 = 0) = \frac{1}{2}$ (why?). Since $u_0 = 1$ we have that $f_2 = u_0 - u_2$. Hence, the formula holds for $n = 1$ as well and this completes the proof of Proposition 3.1.

We have an exact formula for u_{2n}. However, it will be useful to have an estimate for u_{2n} that does not involve binomial coefficients. This is what we do next.

Lemma 3.2. *We have that*

$$u_{2n} = P(S_{2n} = 0 | S_0 = 0) = \frac{1}{2^{2n}} \binom{2n}{n} \sim \frac{1}{\sqrt{\pi n}},$$

where $a_n \sim b_n$ indicates that $\frac{a_n}{b_n}$ converges to 1 as n goes to infinity.

Proof of Lemma 3.2. The proof is a simple application of Stirling's formula:

$$n! \sim \sqrt{2\pi} n^{n+\frac{1}{2}} e^{-n}.$$

For a proof of this formula see, for instance, Schinazi (2011).

By definition of a binomial coefficient we have

$$\binom{2n}{n} = \frac{(2n)!}{n! n!}.$$

By Stirling's formula

$$\binom{2n}{n} \sim \frac{\sqrt{2\pi}(2n)^{2n+\frac{1}{2}} e^{-2n}}{(\sqrt{2\pi} n^{n+\frac{1}{2}} e^{-n})^2}.$$

After simplification of the ratio we get

$$\binom{2n}{n} \sim \frac{2^{2n}}{\sqrt{\pi n}}$$

and hence

$$u_{2n} \sim \frac{1}{\sqrt{\pi n}}.$$

This completes the proof of Lemma 3.2.

Proposition 3.1 and Lemma 3.2 are useful to prove the following important result.

Theorem 3.1. *Starting at 0 the probability that the symmetric random walk will return to 0 is 1. The symmetric random walk is said to be recurrent.*

Proof of Theorem 3.1. We want to show that with probability 1 there is a finite (random) time at which the walk returns to 0. It is obvious that the walk may return to 0 with some positive probability (why?). What is not obvious is that the walk will return to 0 with probability 1.

Let F be the time of the first return to 0. That is,

$$\{F = 2n\} = \{S_2 \neq 0, S_4 \neq 0, \dots, S_{2n-2} \neq 0, S_{2n} = 0\}.$$

In particular,

$$P(F = 2n) = f_{2n}.$$

The event "the random walk will return to 0" is the same as the event $\{F < \infty\}$. Now

$$P(F < \infty) = \sum_{n=1}^{\infty} P(F = 2n) = \sum_{n=1}^{\infty} f_{2n}.$$

To prove the Theorem it is necessary and sufficient to show that this infinite series is 1. We first look at the partial sum. Let

$$a_{2k} = \sum_{n=1}^{2k} f_{2n}.$$

By Proposition 3.1 we have

$$a_{2k} = \sum_{n=1}^{2k} (u_{2n-2} - u_{2n}).$$

All the terms of the sum cancel except for the first and last. Hence,

$$a_{2k} = u_0 - u_{2k}.$$

By Lemma 3.2

$$\lim_{k \to \infty} u_{2k} = 0.$$

Since $u_0 = 1$ we get that

$$\lim_{k \to \infty} a_{2k} = 1.$$

We may conclude that the series $\sum_{n=1}^{\infty} f_{2n}$ converges to 1. This completes the proof of Theorem 3.1.

We finish this section with a rather intriguing result.

Theorem 3.2. *Starting at 0 the expected time for the symmetric random walk to return to 0 is infinite.*

Theorem 3.1 tells us that a return to 0 happens with probability 1. However, Theorem 3.2 tells us that the expected time to return is infinite! The symmetric random walk is said to be *null recurrent*. Theorem 3.2 shows that it may take a very long time for the walk to return.

Proof of Theorem 3.2. With the notation of the proof of Theorem 3.1 we want to prove that the expected value of F is infinite. We have

$$E(F) = \sum_{n=1}^{\infty} (2n) f_{2n}.$$

From Proposition 3.1 and the exact formula for u_{2n} it is not difficult to show (see the problems) that for $n \geq 1$

$$f_{2n} = \frac{1}{2n-1} u_{2n}.$$

Therefore, by Lemma 3.2

$$f_{2n} \sim \frac{1}{2n-1} \frac{1}{\sqrt{\pi n}} \sim \frac{1}{2\sqrt{\pi} n^{\frac{3}{2}}}.$$

Hence,

$$(2n) f_{2n} \sim \frac{1}{\sqrt{\pi} n^{\frac{1}{2}}}.$$

Therefore, the series $\sum_{n=1}^{\infty} (2n) f_{2n}$ diverges (why?) and $E(F) = +\infty$. This completes the proof of Theorem 3.2.

4   Last Return to the Origin

Starting the random walk at 0 let L_{2n} be the time of the last return to the origin up to time $2n$. That is, for $k = 0, 1, \ldots, n$

$$P(L_{2n} = 2k \,|\, S_0 = 0) = P(S_{2k} = 0, S_{2k+2} \neq 0, \ldots, S_{2n} \neq 0 \,|\, S_0 = 0).$$

The next result gives the distribution of this random variable.

Theorem 4.1. *Let L_{2n} be the time of the random walk last return to the origin up to time $2n$. We have for $k = 0, 1, \ldots, n$*

$$P(L_{2n} = 2k \,|\, S_0 = 0) = u_{2k} u_{2n-2k}$$

where

$$u_{2n} = \frac{1}{2^{2n}} \binom{2n}{n}$$

This result is interesting in several ways.

1. Note that the distribution is symmetric in the sense that

$$P(L_{2n} = 2k | S_0 = 0) = P(L_{2n} = 2n - 2k | S_0 = 0).$$

   In particular, assuming n is odd

$$P(L_{2n} < n) = P(L_{2n} > n) = \frac{1}{2}.$$

   From a gambler's perspective this means that there is a probability 1/2 of no equalization for the whole second half of the game, regardless of the length of the game!

2. The most likely values for L_{2n} are the extreme values 0 and $2n$. The least likely value is the mid value (i.e., n if n is even, $n - 1$ and $n + 1$ if n is odd). See the problems. From a gambler's perspective this shows that long leads in the game (i.e., one gambler is ahead for a long time) are more likely than frequent changes in the lead.

3. Theorem 4.1 is called the *discrete arcsine law* because the distribution of L_{2n} can be approximated using the inverse function of sine. See Feller (1968) or Lesigne (2005).

Proof of Theorem 4.1. Since L_{2n} is the time of last return we have

$$P(L_{2n} = 2k | S_0 = 0) = P(S_{2k} = 0, S_{2k+2} \neq 0, \ldots, S_{2n} \neq 0 | S_0 = 0).$$

By the definition of conditional probability we have

$$P(L_{2n} = 2k | S_0 = 0) = P(S_{2k+2} \neq 0, \ldots, S_{2n} \neq 0 | S_0 = 0, S_{2k} = 0) P(S_{2k} = 0 | S_0 = 0).$$

By definition of u_{2n} we have

$$P(S_{2k} = 0 | S_0 = 0) = u_{2k}.$$

To complete the proof we need to show that

$$P(S_{2k+2} \neq 0, \ldots, S_{2n} \neq 0 | S_0 = 0, S_{2k} = 0) = u_{2n-2k}.$$

Given that $S_{2k} = 0$ the information $S_0 = 0$ is irrelevant for the process S_n for $n \geq 2k$ (this is the Markov property). Hence,

$$P(S_{2k+2} \neq 0, \ldots, S_{2n} \neq 0 | S_0=0, S_{2k}=0)=P(S_{2k+2} \neq 0, \ldots, S_{2n} \neq 0 | S_{2k}=0)$$

and by shifting the time by $2k$ we get

$$P(S_{2k+2} \neq 0, \ldots, S_{2n} \neq 0 | S_{2k} = 0) = P(S_2 \neq 0, \ldots, S_{2n-2k} \neq 0 | S_0 = 0).$$

By Proposition 2.2 we have

$$P(S_2 \neq 0, \ldots, S_{2n-2k} \neq 0 | S_0 = 0) = u_{2n-2k}.$$

This completes the proof of Theorem 4.1.

Problems

1. (a) For integers n and k such that $0 \le k \le n$ show that $\binom{n}{k}$ is equal to $\binom{n}{n-k}$.

(b) With the notation of Proposition 1.1 show that $\binom{n}{n^+}$ is equal to $\binom{n}{n^-}$

2. Show that

$$\binom{n-1}{\frac{n+s}{2}-1} - \binom{n-1}{\frac{n+s}{2}} = \frac{s}{n}\binom{n}{\frac{n+s}{2}}$$

3. Show that

$$\frac{1}{2^{2n-1}}\binom{2n-1}{n} = \frac{1}{2^{2n}}\binom{2n}{n}$$

4. Let

$$A = \{S_0 = 0, S_1 > 0, S_2 > 0, \ldots, S_{2n} > 0\}$$

and

$$B = \{S_0 = 0, S_1 < 0, S_2 < 0, \ldots, S_{2n} < 0\}.$$

(a) Show that there are as many paths in A as there are in B.
(b) Show that

$$P(S_0 = 0, S_1 \neq 0, \ldots, S_{2n} \neq 0) = P(A) + P(B).$$

5. (a) Show that the number of paths from $(0,0)$ to $(2n, 2k)$ that do intersect the $s = -1$ axis equals the total number of paths from $(0, -2)$ to $(2n, 2k)$.

(b) Show that the total number of paths from $(0, -2)$ to $(2n, 2k)$ is the same as the total number of paths from $(0, 0)$ to $(2n, 2k + 2)$.

6. Recall that u_{2n} is the probability of a return to 0 at time $2n$ and that

$$u_{2n} = P(S_{2n} = 0 | S_0 = 0) = \frac{1}{2^{2n}} \binom{2n}{n}$$

(a) Compute u_{2n} for $n = 0, 1, \ldots, 10$.

(b) Compare the exact computations done in (a) to the approximation done in Lemma 3.2.

(c) Show that u_{2n} is a decreasing sequence.

(d) Show that $\lim_{n \to \infty} u_{2n} = 0$.

(e) Show that

$$\sum_{n=0}^{\infty} u_{2n} = +\infty.$$

7. (a) Show that

$$P(S_2 = 0 | S_0 = 0) = \frac{1}{2}.$$

(b) Theorem 3.1 states that starting at 0 the symmetric random walk returns to 0 with probability 1. Without using this result show that the probability of returning to 0 is at least 1/2. (Use (a)).

8. (a) Use that $f_{2n} = u_{2n-2} - u_{2n}$ and that

$$u_{2n} = \frac{1}{2^{2n}} \binom{2n}{n}$$

to show that

$$f_{2n} = \frac{1}{2n - 1} u_{2n},$$

for $n \geq 1$.

(b) For which n do we have $u_{2n} = f_{2n}$?

9. Let U_n be a sequence of i.i.d. uniform random variables on $(0, 1)$. Let $S_0 = 0$ and for $n \geq 1$ let

$$S_n = S_{n-1} + 2(U_n < \frac{1}{2}) - 1,$$

where $(U_n < \frac{1}{2}) = 1$ if $U_n < \frac{1}{2}$ and $(U_n < \frac{1}{2}) = 0$ if $U_n > \frac{1}{2}$.

(a) Explain why the algorithm above simulates a random walk.
(b) Use the algorithm to simulate a random walk with 10,000 steps.

10. (a) Use the algorithm in 9 to simulate the first return time to 0 (denoted by F).
(b) Show that F is a very dispersed random variable.

11. Recall that

$$u_{2n} = \frac{1}{2^{2n}} \binom{2n}{n}$$

(a) Compute

$$a_k = u_{2k} u_{10-2k}$$

for $k = 0, 1, \ldots, 10$.
(b) For which k is a_k maximum? Minimum?

12. Recall that L_{2n} is the time of the random walk last return to the origin up to time $2n$. Let n be odd.
Show that

$$P(L_{2n} < n) = P(L_{2n} > n) = \frac{1}{2}.$$

13. I am betting 1\$ on heads for 100 bets.

(a) What is the probability that I am ahead during the whole game?
(b) What is the probability that I am never ahead?

14. Let n and k be positive integers such that $0 \le k \le n$.

(a) Show that

$$P(S_{2n} = 0|S_0 = 0) \ge P(S_{2n} = 0, S_{2k} = 0|S_0 = 0).$$

(b) Show that

$$P(S_{2n} = 0, S_{2k} = 0|S_0 = 0) = P(S_{2n-2k} = 0|S_0 = 0)P(S_{2k} = 0|S_0 = 0).$$

(c) Use (a) and (b) to show that

$$u_{2n} \ge u_{2n-2k} u_{2k}.$$

15. I place $1 bets 30 times in a fair game.

(a) What is the probability that I make a gain of $6?

(b) What is the probability that I make a gain of $6 and that at some point in the game I was behind by $5 or more?

Notes

We followed Feller (1968) in this chapter. There are many more results for random walks there. A more advanced treatment that concentrates on limit theorems for random walks is Lesigne (2005). An advanced beautiful book on random walks is Spitzer (1976).

References

Feller, W.: An Introduction to Probability Theory and Its Applications, vol. I, 3rd edn. Wiley, New York (1968)

Lesigne, E.: Heads or Tails: An Introduction to Limit Theorems in Probability. American Mathematical Society, Providence (2005)

Schinazi, R.B.: From Calculus to Analysis. Birkhauser, Boston (2011)

Spitzer, F.: Principles of Random Walks. Springer, New York (1976)

Chapter 4
Asymmetric and Higher Dimension Random Walks

In this chapter we extend our study of random walks to more general cases. We state an important criterion for recurrence that we apply to random walks.

1 Transience of Asymmetric Random Walks

We have seen in the preceding chapter that the symmetric random walk on $\mathbf{Z}$ is recurrent. That is, a walk starting at the origin will return to the origin with probability 1. We computed the probability of return and we showed that this probability is 1. In many cases it is easier to apply the following recurrence criterion rather than compute the return probability.

First we recall the notation from the preceding chapter.

Let S_n be the position of the walker at time $n \geq 0$ and let

$$u_{2n} = P(S_{2n} = 0 | S_0 = 0).$$

Theorem 1.1. *The random walk* $(S_n)_{n \geq 0}$ *is recurrent if and only if*

$$\sum_{n=0}^{\infty} u_{2n} = +\infty.$$

In fact Theorem 1.1 is a general result that applies to all Markov chains. We will prove it in that context in the next chapter.

Example 1.1. Recall that in the symmetric case we have

$$u_{2n} = P(S_{2n} = 0 | S_0 = 0) = \frac{1}{2^{2n}} \binom{2n}{n} \sim \frac{1}{\sqrt{\pi n}}.$$

© Springer Science+Business Media New York 2014
R.B. Schinazi, *Classical and Spatial Stochastic Processes: With Applications to Biology*, DOI 10.1007/978-1-4939-1869-0_4

Hence, the series $\sum_{n=0}^{\infty} u_{2n}$ diverges. This is another proof that the one dimensional symmetric random walk is recurrent.

Consider the one dimensional random walk $(S_n)_{n \geq 0}$ on $\mathbf{Z}$. We have $S_n = S_{n-1} + 1$ with probability p and $S_n = S_{n-1} - 1$ with probability $q = 1 - p$.

We will now show that if $p \neq \frac{1}{2}$ the random walk is transient. We use Theorem 1.1 to determine for which values of $p \in [0, 1]$ the one dimensional random walk is recurrent.

We take p to be in $(0,1)$. It is easy to see that in order for the walk to come back to the origin at time $2n$ we need n steps to the right and n steps to the left. We have to count all possible combinations of left and right steps. Thus,

$$u_{2n} = \binom{2n}{n} p^n q^n = \frac{(2n)!}{n!n!} p^n q^n.$$

In order to estimate the preceding probability we use Stirling's formula:

$$n! \sim n^n e^{-n} \sqrt{2\pi n}$$

where $a_n \sim b_n$ means that $\lim_{n \to \infty} \frac{a_n}{b_n} = 1$. Stirling's formula yields

$$u_{2n} \sim \frac{(4pq)^n}{\sqrt{\pi n}}.$$

It is easy to check that $4pq < 1$ if $p \neq \frac{1}{2}$ and $4pq = 1$ if $p = \frac{1}{2}$.

Assume first that $p \neq \frac{1}{2}$. By the ratio test from Calculus we see that

$$\sum_{n \geq 0} u_{2n}$$

converges. Hence, the random walk is transient.

The one dimensional simple random walk is recurrent if and only if it is symmetric. In other words, in the asymmetric case there is a strictly positive probability of never returning to the origin.

2 Random Walks in Higher Dimensions

2.1 The Two Dimensional Walk

We consider the two dimensional simple symmetric random walk. The walk is on $\mathbf{Z}^2$ and at each step four transitions (two vertical transitions and two horizontal transitions) are possible. Each transition has probability $1/4$. For instance, starting at $(0,0)$ the walk can jump to $(0,1)$, $(0,-1)$, $(1,0)$ or $(-1,0)$.

In order for the two dimensional random walk to return to the origin in $2n$ steps it must move i units to the right, i units to the left, j units up, and j units down where $2i + 2j = 2n$. Hence

$$u_{2n} = \sum_{i=0}^{n} \frac{(2n)!}{i!i!(n-i)!(n-i)!} (\frac{1}{4})^{2n}.$$

Dividing and multiplying by $(n!)^2$ yields

$$u_{2n} = \frac{(2n)!}{n!n!} \sum_{i=0}^{n} \frac{n!n!}{i!i!(n-i)!(n-i)!} (\frac{1}{4})^{2n},$$

hence

$$u_{2n} = \binom{2n}{n} \sum_{i=0}^{n} \binom{n}{i} \binom{n}{n-i} (\frac{1}{4})^{2n}.$$

We have the following combinatorial identity.

$$\sum_{i=0}^{n} \binom{n}{i} \binom{n}{n-i} = \binom{2n}{n}.$$

See the problems for a proof. Hence,

$$u_{2n}(O, O) = \binom{2n}{n}^2 (\frac{1}{4})^{2n}.$$

Using Stirling's formula we get

$$u_{2n}(O, O) \sim \frac{1}{\pi n}.$$

By Theorem 1.1 this proves that the two dimensional random walk is recurrent (why?).

2.2 The Three Dimensional Walk

We now turn to the analysis of the three dimensional simple symmetric random walk. The walk is on $\mathbf{Z}^3$ and the only transitions that are allowed are plus or minus one unit for one coordinate at the time, the probability of each of these six transitions is $\frac{1}{6}$. By a reasoning similar to the one we did in two dimensions we get

$$u_{2n} = \sum_{i=0}^{n} \sum_{j=0}^{n-i} \frac{(2n)!}{i!i!j!j!(n-i-j)!(n-i-j)!} (\frac{1}{6})^{2n}.$$

We multiply and divide by $(n!)^2$ to get

$$u_{2n} = (\frac{1}{2})^{2n} \binom{2n}{n} \sum_{i=0}^{n} \sum_{j=0}^{n-i} (\frac{n!}{i!j!(n-i-j)!})^2 (\frac{1}{3})^{2n}.$$

Let

$$c(i, j) = \frac{n!}{i!j!(n-i-j)!} \quad \text{for } 0 \le i + j \le n,$$

and

$$m_n = \max_{i,j,0 \le i+j \le n} c(i, j).$$

We have

$$u_{2n} \le \binom{2n}{n} (\frac{1}{2})^{2n} (\frac{1}{3})^n m_n \sum_{i=0}^{n} \sum_{j=0}^{n-i} \frac{n!}{i!j!(n-i-j)!} (\frac{1}{3})^n,$$

but

$$(\frac{1}{3} + \frac{1}{3} + \frac{1}{3})^n = \sum_{i=0}^{n} \sum_{j=0}^{n-i} \frac{n!}{i!j!(n-i-j)!} (\frac{1}{3})^n = 1.$$

Hence

$$u_{2n} \le (\frac{1}{2})^{2n} \binom{2n}{n} (\frac{1}{3})^n m_n.$$

We now need to estimate m_n. Suppose that the maximum of the $c(i, j)$ occurs at (i_0, j_0) then the following inequalities must hold:

$$c(i_0, j_0) \ge c(i_0 - 1, j_0)$$
$$c(i_0, j_0) \ge c(i_0 + 1, j_0)$$
$$c(i_0, j_0) \ge c(i_0, j_0 - 1)$$
$$c(i_0, j_0) \ge c(i_0, j_0 + 1).$$

These inequalities imply that

$$n - i_0 - 1 \le 2j_0 \le n - i_0 + 1$$
$$n - j_0 - 1 \le 2i_0 \le n - j_0 + 1.$$

Hence,

$$n - 1 \le 2j_0 + i_0 \le n + 1$$
$$n - 1 \le 2i_0 + j_0 \le n + 1.$$

Therefore, the approximate values for i_0 and j_0 are $n/3$. We use this to get

$$u_{2n} \le (\frac{1}{2})^{2n} \binom{2n}{n} c(\frac{n}{3}, \frac{n}{3})(\frac{1}{3})^n.$$

We use again Stirling's formula to get that the right-hand side of the last inequality is asymptotic to $\frac{C}{n^{3/2}}$ where C is a constant. This proves that $\sum_{n \ge 0} u_{2n}$ is convergent and therefore the three dimensional random walk is transient. In other words there is a positive probability that the three dimensional random walk will never return to the origin. This is in sharp contrast with what happens for the random walk in dimensions one and two.

3   The Ruin Problem

Consider a gambler who is making a series of 1\$ bets. He wins \$1 with probability p and loses \$1 with probability $q = 1 - p$. His initial fortune is m. He plays until his fortune is 0 (he is ruined) or N. We would like to compute the probability of ruin. Note that

$$P(\text{ruin}) = P(F_0 < F_N | X_0 = m),$$

where F_i is the time of the *first* visit to state i. That is,

$$F_i = \min\{n \ge 1 : X_n = i\}.$$

The following result gives explicit formulas for the probability of ruin.

Proposition 3.1. *Assume that the initial fortune of the gambler is m and that he plays until he is ruined or his fortune is $N > m$. For $p \ne q$ the probability of ruin is*

$$P(\text{ruin}) = \frac{(\frac{q}{p})^m - (\frac{q}{p})^N}{1 - (\frac{q}{p})^N}.$$

For $p = q = \frac{1}{2}$ we have

$$P(\text{ruin}) = 1 - \frac{m}{N}.$$

Proof of Proposition 3.1. Let $0 < m < N$ and let

$$u(m) = P(F_0 < F_N | X_0 = m).$$

That is, $u(m)$ is the probability of ruin if the starting fortune is m. Set $u(0) = 1$ and $u(N) = 0$. We condition on the first bet to get

$$u(m) = P(F_0 < F_N; X_1 = m - 1 | X_0 = m) + P(F_0 < F_N; X_1 = m + 1 | X_0 = m).$$

By the Markov property this implies that for $1 \le m < N$

$$u(m) = qu(m - 1) + pu(m + 1) \tag{3.1}$$

Equation (3.1) for $m = 1, \ldots, N - 1$ are called difference equations. We look for r such that $u(m) = r^m$ is a solution of (3.1). Therefore,

$$r^m = qr^{m-1} + pr^{m+1}.$$

Hence, r must be a solution of

$$r = q + pr^2.$$

This quadratic equation is easily solved and we see that $r = 1$ or $r = \frac{q}{p}$. There are two cases to consider.

- If $\frac{q}{p} \ne 1$, then we let

$$u(m) = A(1)^m + B(\frac{q}{p})^m = A + B(\frac{q}{p})^m.$$

It is easy to check that $A + B(\frac{q}{p})^m$ is a solution of (3.1). It is also possible to show that $A + B(\frac{q}{p})^m$ is the unique solution of (3.1) under the boundary conditions $u(0) = 1$ and $u(N) = 0$, see the problems. This is important for we want to be sure that $u(m)$ is the ruin probability, not some other solution of (3.1). We now find A and B by using the boundary conditions. We get

$$u(0) = A + B = 1$$

$$u(N) = A + B(\frac{q}{p})^N = 0.$$

The solution of this system of linear equations is

$$A = -\frac{(\frac{q}{p})^N}{1 - (\frac{q}{p})^N}, B = \frac{1}{1 - (\frac{q}{p})^N}.$$

Hence, for $\frac{q}{p} \neq 1$ and $1 < m < N$

$$u(m) = A + B(\frac{q}{p})^m = \frac{(\frac{q}{p})^m - (\frac{q}{p})^N}{1 - (\frac{q}{p})^N}.$$

- If $\frac{q}{p} = 1$, then the equation $r = q + pr^2$ has a double root $r = 1$. In this case we let

$$u(m) = A + Bm.$$

Again we can check that $A + Bm$ is the unique solution of (3.1) under the conditions $u(0) = 1$ and $u(N) = 0$, see the Problems. We find A and B using the boundary conditions. From $u(0) = 1$ we get $A = 1$ and from $u(N) = 0$ we get $B = -\frac{1}{N}$. Hence, for $\frac{q}{p} = 1$ and $1 < m < N$ we have

$$u(m) = 1 - \frac{m}{N}.$$

This completes the proof of Proposition 3.1.

Example 3.1. A roulette has 38 pockets, 18 are red, 18 are black, and 2 are green. A gambler bets on red at the roulette. The gambler bets 1\$ at a time. His initial fortune is 90\$ and he plays until his fortune is 0\$ or 100\$. Note that $p = \frac{18}{38}$ and so $\frac{q}{p} = \frac{20}{18}$. By Proposition 3.1

$$P(\text{ruin}) = P(F_0 < F_{100}|X_0 = 90) = \frac{(\frac{20}{18})^{90} - (\frac{20}{18})^{100}}{1 - (\frac{20}{18})^{100}} \sim 0.65.$$

Hence, even though the player is willing to lose \$90 to gain \$10 his probability of succeeding is only about 1/3. This is so because the game is unfair.

Example 3.2. We continue Example 3.1. What if the game is fair (i.e., $p = q$)?
This time the probability of ruin is

$$P(\text{ruin}) = P(F_0 < F_{100}|X_0 = 90) = 1 - \frac{90}{100} = \frac{1}{10}.$$

So in a fair game the probability of success for the player goes up to 90 %.

Example 3.3. Going back to Example 3.1. What if the gambler plays 10$ a bet instead of 1$?

Our ruin probability formula holds for games that go up or down one unit at a time. To use the formula we define one unit to be $10. Hence,

$$P(\text{ruin}) = P(F_0 < F_{10} | X_0 = 9) = \frac{(\frac{20}{18})^9 - (\frac{20}{18})^{10}}{1 - (\frac{20}{18})^{10}} \sim 0.15.$$

Therefore, this is a much better strategy. The probability of success goes to 85 % if the stakes are 10$ instead of 1$. However, this is still an uphill battle for the gambler. He makes 10$ with probability 85 % but loses 90$ with probability 15 %. Hence his average gains are

$$10 \times 0.85 - 90 \times 0.15 = -5\$.$$

That is, the gambler is expected to lose $5.

3.1   The Greedy Gambler

Assume now that our gambler is greedy. He is never happy with his gains and goes on playing forever. What is his probability of ruin?

Recall that for $p \neq q$ we have

$$P(\text{ruin}) = P(F_0 < F_N | X_0 = m) = \frac{(\frac{q}{p})^m - (\frac{q}{p})^N}{1 - (\frac{q}{p})^N}.$$

Since the gambler will never stop unless he is ruined we let N go to infinity in the formula. There are three cases.

- Assume that $p > q$ (the gambler has an advantage). Since $(\frac{q}{p})^N$ goes to 0 as N goes to infinity we have

$$P(\text{ruin}) \to (\frac{q}{p})^m.$$

Hence, ruin is not certain and the probability of ruin goes down as the initial fortune m goes up.

- Assume that $p < q$ (the gambler has a disadvantage). Since $(\frac{q}{p})^N$ goes to infinity with N

$$P(\text{ruin}) \sim \frac{-(\frac{q}{p})^N}{-(\frac{q}{p})^N} = 1.$$

Hence, ruin is certain in this case.

- Assume that $p = q = \frac{1}{2}$. In this case ruin is certain as well. See the problems.

3.2 Random Walk Interpretation

We can also interpret these results using random walks. Assume that a random walk starts at some $m > 0$. What is the probability that the random walk eventually reaches 0?

Note that the probability of eventually reaching 0 is exactly the probability of ruin. Hence, we have

- Assume that $p > q$. Then the probability that the random walk eventually reaches 0 is $(\frac{q}{p})^m$. Therefore, the probability of never reaching 0 is $1 - (\frac{q}{p})^m < 1$. This is not really surprising since the random walk has a bias towards the right.
- Assume that $p < q$. Then the probability that the random walk eventually reaches 0 is 1. In this case the walk is biased towards the left and will eventually reach 0.

3.3 Duration of the Game

We continue working on the ruin question. We now compute the expected time the game lasts. Starting with a fortune equal to m the gambler plays until his fortune is 0 or N. We are interested in D_m the expected duration of the game.

Proposition 3.2. *Assume that the initial fortune of the gambler is m and that he plays until he is ruined or his fortune is $N > m$. Let D_m be the expected duration of the game. For $p \neq q$*

$$D_m = \frac{m}{1 - 2p} - \frac{N}{1 - 2p} \frac{1 - (\frac{q}{p})^m}{1 - (\frac{q}{p})^N}.$$

For $p = q$

$$D_m = m(N - m).$$

Proof of Proposition 3.2. We will use the same technique used for Proposition 3.1.

We set $D_0 = D_N = 0$. Assume from now that $0 < m < N$. Condition on the outcome of the first bet. If the first bet is a win, then the chain is at $m + 1$ after one step and the expected remaining number of steps for the chain to be absorbed (at 0 or N) is D_{m+1}. If the first bet is a loss, then the chain is at $m - 1$ after one step and the expected remaining number of steps for the chain to be absorbed is D_{m-1}. Thus,

$$D_m = p(1 + D_{m+1}) + q(1 + D_{m-1}).$$

Hence, for $0 < m < N$ we have

$$D_m = 1 + pD_{m+1} + qD_{m-1} \tag{3.2}$$

In order to solve these equations, it is convenient to get rid of the 1 in (3.2). To do so we set $a_m = D_m - D_{m-1}$ for $m \geq 1$. Therefore,

$$a_m = pa_{m+1} + qa_{m-1}. \tag{3.3}$$

We look for solutions of the type $a_i = r^i$ for a constant r. Setting $a_i = r^i$ in (3.3) gives the following characteristic equation.

$$r = pr^2 + q.$$

Note this is the same equation we had in proof of Proposition 3.1. The characteristic equation has two solutions 1 and $\frac{q}{p}$. As before it is important to know whether these roots are distinct. We will treat the case $p \neq q$ and leave the case $p = q$ for the problems.

The unique solution of (3.3) is

$$a_m = A + B(\frac{q}{p})^m$$

where A and B will be determined by the boundary conditions.

We sum the preceding equality for $j = 1$ to m to get

$$\sum_{j=1}^{m} a_j = \sum_{j=1}^{m} (D_j - D_{j-1}) = \sum_{j=1}^{m} (A + B(\frac{q}{p})^j).$$

This yields

$$D_i - D_0 = D_i = iA + \frac{q}{p} \frac{1 - (\frac{q}{p})^i}{1 - \frac{q}{p}} B \text{ for } i \geq 1 \tag{3.4}$$

We now use boundary conditions to compute A and B. Writing $D_N = 0$ yields

$$NA + \frac{q}{p} \frac{1 - (\frac{q}{p})^N}{1 - \frac{q}{p}} B = 0. \tag{3.5}$$

By (3.2) $D_1 = 1 + pD_2$. Using (3.4) to substitute D_1 and D_2 we get

$$A + \frac{q}{p}B = 1 + p(2A + \frac{q}{p}\frac{1 - (\frac{q}{p})^2}{1 - \frac{q}{p}}B). \tag{3.6}$$

Note that (3.5) and (3.6) is a system of two linear equations in A and B. The solution is given by

$$A = \frac{1}{1 - 2p} \quad \text{and } B = \frac{1}{\frac{q}{N}(1 - (\frac{q}{p})^N)}.$$

We use these values in (3.5) to get for $m \geq 1$

$$D_m = \frac{m}{1 - 2p} - \frac{N}{1 - 2p}\frac{1 - (\frac{q}{p})^m}{1 - (\frac{q}{p})^N}.$$

This completes the proof of Proposition 3.2 for the case $p \neq q$. The case $p = q$ is treated in a similar fashion. See the problems.

The numerical values of D_m turn out to be quite large.

Example 3.4. We go back to Example 3.1. Hence, $\frac{q}{p} = \frac{20}{18}$, $N = 100\$$ and the initial fortune is $m = 90\$$. By Proposition 3.2 we get

$$D_{90} = \frac{90}{1 - 2\frac{9}{19}} - \frac{100}{1 - 2\frac{9}{19}}\frac{1 - (\frac{20}{18})^{90}}{1 - (\frac{20}{18})^{100}} \sim 1,048.$$

That is, the game is expected to last $1,048 steps. This is unexpectedly long! After all the player will stop if he makes 10\$ or loses his 90\$.

Example 3.5. We use the same values as in Example 3.5 with the difference that $p = q$. We get by Proposition 3.2

$$D_{90} = 90(100 - 90) = 900.$$

Again this is quite long. We need an average of 900 bets to gain $10 or lose $90.

Problems

1. A gambler makes a series of one dollar bets. He decides to stop playing as soon as he wins 40\$ or he loses 10\$.

(a) Assume this is a fair game, compute the probability that the player wins.

(b) Assume this is a slightly unfair game:

$$p = 9/19 \qquad q = 10/19,$$

and compute the probability that the player wins.

(c) Compute the expected gains in (a) and (b).

2. My initial fortune is \$980. I stop playing when I reach \$0 or \$1,000. Assume that the probability of winning a bet is $p = \frac{18}{38}$.

(a) Assume that I bet \$1 at a time. Compute the probability of ruin.
(b) Compute the probability of ruin if I bet \$20 at a time.
(c) I use the strategy in (b) once a year. What is the probability that I win 10 years in a row?

3. Show that if the game is fair (i.e., $p = 1/2$) the probability of ruin does not change if the stakes are changed.

4. Assume that $p = 1/2$ and let N go to infinity in the ruin probability. What is the limit of the ruin probability? Interpret the result.

5. Consider the following (hypothetical) slot machine. Bets are 1\$ each. The player wins 1\$ with probability 0.49 and loses 1\$ with probability 0.51. Assume that the slot machine starts with \$100.

(a) What is the probability that the slot machine eventually runs out of money?
(b) Same question as (a) if the stakes are 10\$.
(c) With 1\$ bets, what is the probability that at least one of 20 machines eventually runs out of money?

6. A gambler makes a series of one dollar bets. He decides to stop playing as soon as he wins 40\$ or he loses 10\$.

(a) Assume this is a fair game, compute the expected number of bets before the game stops.
(b) Assume this is a slightly unfair game with $p = 9/19$. Compute the expected number of bets before the game stops.

7. Assume that the probability of winning a bet is $p = \frac{9}{19}$, the initial fortune is $m = 90\$$ and $N = 100\$$. The gambler plays until he is ruined or his fortune is N. Simulate the duration of the game 10 times.

8. Using the same values as in Problem 7 except that $p = \frac{1}{2}$ do 10 simulations of the duration of the game.

9. Assume that $p \neq q$. Show that $A + B(\frac{q}{p})^m$ is a solution of (3.1).

10. Assume that $p \neq q$. Assume also that $(b_m)_{m \geq 0}$ is a solution of (3.1).

(a) Show that there are A and B such that

$$b_0 = A + B$$

$$b_1 = A + B\frac{q}{p}$$

(b) Show (by induction) that for all $m \geq 0$

$$b_m = A + B(\frac{q}{p})^m$$

for the constants A and B found in (a).

(c) Conclude that there exist A and B so that $A + B(\frac{q}{p})^m$ is the unique solution of (3.1).

11. Assume that $p = q$. Show that there exist A and B so that $A + Bm$ is the unique solution of (3.1). (Use steps similar to the ones in problems 9 and 10.)

12. Assume that $p = q = \frac{1}{2}$.

(a) Show that all the solutions of (3.3) are of the type $a_m = A + Bm$.

(b) Show that $D_m = Am + \frac{B}{2}m(m + 1)$.

(c) Show that $D_m = m(N - m)$ for $0 \leq m \leq N$.

13. We give a combinatorial proof of the following identity.

$$\binom{2n}{n} = \sum_{i=0}^{n} \binom{n}{i}\binom{n}{n-i}.$$

Assume that we have n black balls (numbered from 1 to n) and m white balls (numbered from 1 to m). The balls are mixed and we pick k balls.

(a) Show that there are

$$\binom{m + n}{k}$$

ways to pick k balls.

(b) By counting the number of black balls among the k balls show that there are

$$\sum_{i=0}^{n} \binom{n}{i}\binom{m}{k-i}$$

ways to select k balls.

(c) Prove the identity.

14. We give another proof of the identity in 14. Write Newton's formula for $(1+t)^n$ and $(1 + t)^{2n}$ in the identity

$$(1 + t)^n (1 + t)^n = (1 + t)^{2n},$$

and identify the coefficients of the polynomials on both sides of the equality.

15. Consider a two dimensional random walk $S_n = (X_n, Y_n)$. Let $S_0 = (0, 0)$. At each step either the first coordinate moves or the second coordinate moves but not both. We define a new stochastic process $(Z_n)_{n \geq 0}$ by only keeping track of the first coordinate when it moves. For instance, if the path of S_n is $(0, 0)$, $(0, 1)$, $(0, 2)$, $(-1, 2)$ and $(-2, 2)$ we set $Z_0 = 0$, $Z_1 = -1$ and $Z_2 = -2$.

(a) Show that $(Z_n)_{n \geq 0}$ is one dimensional random walk.
(b) Use $(Z_n)_{n \geq 0}$ to show that if $(S_n)_{n \geq 0}$ is not symmetric in its first coordinate then $(S_n)_{n \geq 0}$ is transient.
(c) Show that any asymmetric random walk in any dimension is transient.

16. Show that symmetric random walks in dimensions $d \geq 4$ are transient. (Use the method of problem 15.)

Notes and references. See the notes and references of the preceding chapter.

Chapter 5
Discrete Time Markov Chains

Branching processes and random walks are examples of Markov chains. In this chapter we study general properties of Markov chains.

1   Classification of States

We start with two definitions.

Definition 1.1. A discrete time stochastic process is a sequence of random variables $(X_n)_{n \geq 0}$ defined on the same probability space and having values on the same countable space S.

We will take S to be the set of positive integers in most of what we will do in this chapter.

Definition 1.2. A Markov process is a stochastic process for which the future depends on the past and the present only through the present. More precisely, a stochastic process X_n is said to be Markovian if for any states $x_1, x_2, \ldots, x_k, x_n$ in S, any integers $n_1 < n_2 < \ldots < n_k < n$ we have

$$P(X_n = x_n | X_{n_1} = x_1, X_{n_2} = x_2, \ldots, X_{n_k} = x_k) = P(X_n = x_n | X_{n_k} = x_k).$$

We define the one-step transition probability by

$$p(i, j) = P(X_{n+1} = j | X_n = i) \text{ for all } i, j \in S \text{ and all } n \geq 0.$$

Observe that we are assuming that the transition probabilities do not depend on the time variable n, that is, we consider Markov chains with homogeneous transition

© Springer Science+Business Media New York 2014
R.B. Schinazi, *Classical and Spatial Stochastic Processes: With Applications to Biology*, DOI 10.1007/978-1-4939-1869-0_5

probabilities. The $p(i, j)$ must have the following (probability) properties

$$p(i, j) \geq 0 \text{ and } \sum_{j \in S} p(i, j) = 1.$$

We are now going to compute the transition probabilities for two examples.

Example 1.1. Consider a simple random walk $(S_n)_{n \geq 0}$ on $\mathbf{Z}$. If the walk is at i it jumps to $i + 1$ with probability p or to $i - 1$ with probability $q = 1 - p$.

This is a Markov chain since once we know S_n we can compute the distribution of S_{n+1}. We have that $S = \mathbf{Z}$ and the transition probabilities are

$$p(i, i + 1) = p \text{ and } p(i, i - 1) = 1 - p = q \text{ for all i } \in \mathbf{Z}.$$

All other $p(i, j)$ are zero.

Example 1.2. Consider a Bienaymé–Galton–Watson (BGW) process $(Z_n)_{n \geq 0}$. The state space S of $(Z_n)_{n \geq 0}$ is the set of positive (including zero) integers. We suppose that each individual gives birth to Y particles in the next generation where Y is a positive integer-valued random variable with distribution $(p_k)_{k \geq 0}$. In other words

$$P(Y = k) = p_k, \text{ for } k = 0, 1, \ldots.$$

Observe that

$$p(i, j) = P(Z_{n+1} = j | Z_n = i) = P(\sum_{k=1}^{i} Y_k = j) \text{ for } i \geq 1, j \geq 0,$$

where $(Y_k)_{1 \leq k \leq i}$ is a sequence of independent identically distributed (i.i.d.) random variables with distribution $(p_k)_{k \geq 0}$. This shows that the distribution of Z_{n+1} can be computed using the distribution of Z_n only. Hence, $(Z_n)_{n \geq 0}$ has the Markov property.

Define the n-steps transition probability by

$$p_n(i, j) = P(X_{n+m} = j | X_m = i) \text{ for all } i, j \in S \text{ and all } m \geq 0.$$

In particular, $p_1(i, j) = p(i, j)$. We will set $p_0(i, i) = 1$ and $p_0(i, j) = 0$ for $i \neq j$.

1.1 Decomposition of the Chain

We will show that a Markov Chain can be decomposed in classes. We start with two useful properties.

Proposition 1.1. *For all $n \geq 2$ and $i, i_1, \ldots, i_{n-1}, j$ in S we have*

$$p_n(i, j) \geq p(i, i_1)p(i_1, i_2) \ldots p(i_{n-1}, j).$$

We will not give a formal proof of Proposition 1.1. All it says is that the probability to go from i to j in n steps is larger than the probability to go from i to j using the specific path $i, i_1, i_2, \ldots, i_{n-1}, j$. Moreover, by the Markov property the probability of this specific path is $p(i, i_1)p(i_1, i_2) \ldots p(i_{n-1}, j)$.

Another useful property is the following.

Proposition 1.2. *For all positive integers n and m and for all states i, j, ℓ we have*

$$p_{n+m}(i, j) \geq p_n(i, \ell)p_m(\ell, j).$$

To go from i to j in $n + m$ steps we may go from i to ℓ in n steps and from ℓ to j in m steps. This is one possibility among possibly many others this is why we have an inequality. By the Markov property the probability of going from i to ℓ in n steps and from ℓ to j in m steps is $p_n(i, \ell)p_m(\ell, j)$.

We now turn to the decomposition of a Markov chain.

Definition 1.3. Consider a Markov chain $(X_n)_{n \geq 0}$. We say that two states i and j are in the same class if there are integers $n_1 \geq 0$ and $n_2 \geq 0$ such that $p_{n_1}(i, j) > 0$ and $p_{n_2}(j, i) > 0$. In words, i and j are in the same class if the Markov chain can go from i to j and from j to i in a finite number of steps.

We defined $p_0(i, i) = 1$ for every state i, so every state i is in the same class as itself. Note that we may have classes of one element.

Example 1.3. Consider the following chain on $\{0, 1, 2\}$ with the transition matrix

$$\begin{pmatrix} 0 & 1 & 0 \\ 0 & \frac{1}{2} & \frac{1}{2} \\ 0 & \frac{1}{3} & \frac{2}{3} \end{pmatrix}$$

The first line of the matrix gives the following transition probabilities: $p(0, 0) = 0$, $p(0, 1) = 1$, $p(0, 2) = 0$. The second line gives $p(1, 0) = 0$, $p(1, 1) = \frac{1}{2}$, $p(1, 2) = \frac{1}{2}$.

We see that from state 0 we go to state 1 with probability one. But from state 1 we cannot go back to state 0. So 1 and 0 are in different classes. We have $p(1, 2) = \frac{1}{2}$ and $p(2, 1) = \frac{1}{3}$. Thus, states 1 and 2 are in the same class. We have two classes for this chain: $\{0\}$ and $\{1, 2\}$.

Example 1.4. Consider the random walk on the integers. Take p in $(0, 1)$. Let i, j be two integers and assume that $i < j$. By Proposition 1.1 we have

$$p_{j-i}(i, j) \geq p(i, i + 1)p(i + 1, i + 2) \ldots p(j - 1, j) = p^{j-i} > 0.$$

We also have

$$p_{j-i}(j,i) \geq p(j, j-1)p(j-1, j-2)\ldots p(i+1,i) = q^{j-i} > 0.$$

Thus, all states are in the same class for this Markov chain and the chain is said to be irreducible.

Definition 1.4. A Markov chain is said to be irreducible if all states are in the same class.

Note that the chain in Example 1.3 is not irreducible while the chain in Example 1.4 is.

1.2   A Recurrence Criterion

Definition 1.5. Consider a Markov chain $(X_n)_{n \geq 0}$. We say that state i is recurrent for this chain if starting at i the chain returns to i with probability 1. A non-recurrent state is said to be transient.

The next theorem gives us a recurrence criterion in terms of the $p_n(i,i)$.

Theorem 1.1. *Consider a Markov chain* $(X_n)_{n \geq 0}$. *A state i in* $\mathcal{S}$ *is recurrent if and only if*

$$\sum_{n=0}^{\infty} p_n(i,i) = \infty.$$

Proof of Theorem 1.1. Let $f_k(i,i)$ be the probability that starting at i the chain returns to i for the *first* time at time k. Consider

$$f = \sum_{k=1}^{\infty} f_k(i,i).$$

The number f is the probability that the chain eventually returns to i. This is so because the events {the first return occurs at k} are mutually exclusive for $k = 1, 2, \ldots$. By adding the probabilities of these events we get the probability that the chain will return to i in a finite (random) time. Hence, the state i is recurrent if and only if $f = 1$. This proof will show that $f = 1$ if and only if $\sum_{n=0}^{\infty} p_n(i,i) = \infty$.

We have that

$$p_n(i,i) = \sum_{k=1}^{n} f_k(i,i) p_{n-k}(i,i) \tag{1.1}$$

This is so because if the chain is to return to i at time n then it must return to i for the first time at some time $k \leq n$. If the chain is at i at time k (where k is possibly n), then it has to return to i in $n - k$ steps. By the Markov and homogeneity properties of the chain this probability is $p_{n-k}(i,i)$. Note that since $f_k(i,i)$ is the first return to i events corresponding to different k's are mutually exclusive and we may add the different probabilities. We now use (1.1) to get a relation between moment generating functions. For s in $[0, 1]$ we have for all $n \geq 1$

$$s^n p_n(i,i) = s^n \sum_{k=1}^{n} f_k(i,i) p_{n-k}(i,i).$$

Summing the preceding equality over all $n \geq 1$ yields

$$\sum_{n=1}^{\infty} s^n p_n(i,i) = \sum_{n=1}^{\infty} s^n \sum_{k=1}^{n} f_k(i,i) p_{n-k}(i,i) \tag{1.2}$$

Let U and F be defined by

$$U(s) = \sum_{n=0}^{\infty} s^n p_n(i,i)$$

and

$$\sum_{n=1}^{\infty} s^n f_n(i,i).$$

Note that the sum for U starts at $n = 0$ while it starts at $n = 1$ for F. Recall from Calculus that if the series $\sum_{n=1}^{\infty} a_n$ and $\sum_{n=0}^{\infty} b_n$ are absolutely convergent then

$$\sum_{n=0}^{\infty} a_n \sum_{n=1}^{\infty} b_n = \sum_{n=1}^{\infty} \sum_{k=1}^{n} a_k b_{n-k} \tag{1.3}$$

Applying this fact to the r.h.s. of (1.2) with $a_n = f_n(i,i)s^n$ and $b_n = p_n(i,i)s^n$ we get

$$\sum_{n=1}^{\infty} s^n \sum_{k=1}^{n} f_k(i,i) p_{n-k}(i,i) = \sum_{n=1}^{\infty} \sum_{k=1}^{n} s^k f_k(i,i) s^{n-k} p_{n-k}(i,i) = U(s)F(s).$$

On the other hand, for the l.h.s. of (1.2) we have

$$\sum_{n=1}^{\infty} s^n p_n(i,i) = U(s) - s^0 p_0(i,i) = U(s) - 1.$$

Hence, for s in $[0, 1]$

$$U(s) - 1 = U(s)F(s) \tag{1.4}$$

In order to finish the proof we need the following lemma from Analysis. For a proof see, for instance, Proposition A 1.9 in Port (1994).

Lemma 1.1. *Let $b_n \geq 0$ for all $n \geq 1$. Assume that the series $\sum_{n=1}^{\infty} b_n s^n$ converges for all s in $[0, 1)$ then*

$$\lim_{s \to 1^-} \sum_{n=1}^{\infty} b_n s^n = \sum_{n=1}^{\infty} b_n,$$

where both sides of the equality may be infinite.

We now apply Lemma 1.1 to U and F. By Lemma 1.1

$$\lim_{s \to 1^-} F(s) = \sum_{n=1}^{\infty} f_n(i, i) = f,$$

where f is in $[0, 1]$ (it is a probability). Similarly,

$$\lim_{s \to 1^-} U(s) = \sum_{n=0}^{\infty} p_n(i, i) = u,$$

where u is either a positive number or $+\infty$.

If u is a finite number, then letting $s \to 1^-$ in (1.4) yields

$$u(1 - f) = 1$$

and we have $f < 1$. The state i is transient.

Assume now that $u = +\infty$. By (1.4) we have

$$U(s) = \frac{1}{1 - F(s)}.$$

By letting $s \to 1^-$ we see that since $u = +\infty$ we must have $1 - f = 0$. That is, i is recurrent. This completes the proof of Theorem 1.1.

Next we will give several important consequences and applications of Theorem 1.1.

Corollary 1.1. *All states that are in the same class are either all recurrent or all transient.*

A class of recurrent states will be called a recurrent class. An irreducible chain with recurrent states will be called a recurrent chain.

Proof of Corollary 1.1. Assume that states i and j are in the same class. By definition this means that there are positive integers m and n such that

$$p_n(i, j) > 0 \text{ and } p_m(j, i) > 0.$$

Observe that by Proposition 1.2 we have for any states i, j, ℓ

$$p_{m+n}(i, j) \geq p_m(i, \ell) p_n(\ell, j).$$

We iterate twice the preceding inequality to get

$$p_{m+n+r}(j, j) \geq p_m(j, i) p_r(i, i) p_n(i, j),$$

for any positive integer r. We sum over all r to get

$$\sum_{r=0}^{\infty} p_{m+n+r}(j, j) \geq \sum_{r=0}^{\infty} p_m(j, i) p_r(i, i) p_n(i, j) = p_m(j, i) p_n(i, j) \sum_{r=0}^{\infty} p_r(i, i).$$

Since $p_m(j, i) p_n(i, j) > 0$ if $\sum_{r=0}^{\infty} p_r(i, i)$ diverges so does $\sum_{r=0}^{\infty} p_r(j, j)$ (why?). On the other hand, if $\sum_{r=0}^{\infty} p_r(j, j)$ converges so does $\sum_{r=0}^{\infty} p_r(i, i)$. This completes the proof of Corollary 1.1.

The next result will be helpful in the sequel.

Corollary 1.2. *Let j be a transient state then for all i*

$$\sum_{n=1}^{\infty} p_n(i, j) < \infty.$$

Proof of Corollary 1.2. With the notation of Theorem 1.1 we have

$$p_n(i, j) = \sum_{k=1}^{n} f_k(i, j) p_{n-k}(j, j).$$

In words, if the chain is to go from i to j in n steps it must get to j for the first time at some time $k \leq n$ and then it has $n - k$ steps to get to j. Summing both sides for all $n \geq 0$ we get

$$\sum_{n=0}^{\infty} p_n(i, j) = \sum_{n=0}^{\infty} \sum_{k=1}^{n} f_k(i, j) p_{n-k}(j, j).$$

Observe now that $\sum_{k=1}^{\infty} f_k(i, j)$ is the probability that the chain starting at i eventually gets to j and is therefore less than or equal to 1. By Theorem 1.1 $\sum_{n=0}^{\infty} p_n(j, j)$ converges. Hence, by (1.3)

$$\sum_{n=0}^{\infty}\sum_{k=1}^{n} f_k(i,j) p_{n-k}(j,j) = \sum_{k=1}^{\infty} f_k(i,j) \sum_{n=0}^{\infty} p_n(j,j) < +\infty.$$

This completes the proof of Corollary 1.2.

1.3   Finite Markov Chains

We say that a set C of states is closed if no state inside of C leads to a state outside of C. That is, C is closed if for all i in C and j not in C we have $p(i,j) = 0$.

Example 1.5. Consider a Markov chain with the transition probabilities

$$\begin{pmatrix} 0 & 1 & 0 & 0 & 0 \\ \frac{1}{3} & \frac{1}{3} & \frac{1}{3} & 0 & 0 \\ 0 & 0 & \frac{1}{2} & \frac{1}{2} & 0 \\ 0 & 0 & 0 & \frac{1}{2} & \frac{1}{2} \\ 0 & 0 & 0 & \frac{1}{2} & \frac{1}{2} \end{pmatrix}.$$

Let the states be $S = \{1,2,3,4,5\}$. There are three classes: $\{1,2\}$, $\{3\}$, and $\{4,5\}$. There is only one closed class: $\{4,5\}$. We now give a graphical method to decide whether a class is closed, see Fig. 5.1.

We now examine the relation between closed and recurrent.

Example 1.6. A closed set is not necessarily recurrent.

We give an example. Consider the simple random walk on $\mathbf{Z}$ with $p \neq \frac{1}{2}$. This is an irreducible chain (all states are in the same class). Hence, $\mathbf{Z}$ is a closed class for this chain (why?). But we know that this chain is transient.

As the next result shows a recurrent class has to be closed.

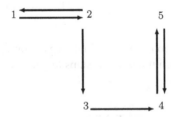

Fig. 5.1 We draw an *arrow* from state i to state j whenever the entry $p(i,j)$ is strictly positive. We see that no *arrow* leaves the class $\{4,5\}$ so this is a closed class. On the other hand, there are *arrows* leaving classes $\{1,2\}$ and $\{3\}$. Hence, these two classes are not closed

Theorem 1.2. *(a) A recurrent class is closed.*
(b) A finite closed class is recurrent.

This is a very useful result for finite chains for it is easy to check whether a class is closed. For instance, going back to Example 1.5, we have that $\{4, 5\}$ is recurrent because it is finite and closed while the two other classes are transient since they are not closed.

Proof of Theorem 1.2. We first prove (a). We prove the contrapositive. Assume that $\mathcal{C}$ is not closed. Therefore, there exist i in $\mathcal{C}$, j not in $\mathcal{C}$ such that $p(i, j) > 0$. Observe that the chain cannot go from j to i. If it could j and i would be in the same class and they are not. Hence, once the chain reaches j it never comes back to i. Therefore,

$$P_i(\text{no return to } i) \geq p(i, j) > 0.$$

Hence, $\mathcal{C}$ is not recurrent (why?). This proves (a).

We now turn to (b).

We do a proof by contradiction. Assume $\mathcal{C}$ is a finite closed transient class. Let i and j be in $\mathcal{C}$. By Corollary 1.2 we have

$$\lim_{n \to \infty} p_n(i, j) = 0,$$

(why?) and so

$$\sum_{j \in \mathcal{C}} \lim_{n \to \infty} p_n(i, j) = 0.$$

Since $\mathcal{C}$ is finite we may exchange the limit and the sum to get (1.4),

$$\lim_{n \to \infty} \sum_{j \in \mathcal{C}} p_n(i, j) = 0,$$

but $\sum_{j \in \mathcal{C}} p_n(i, j) = P(X_n \in \mathcal{C} | X_0 = i)$. Since $\mathcal{C}$ is closed the preceding probability is one. This contradicts (1.4). Therefore $\mathcal{C}$ must be recurrent and this completes the proof of Theorem 1.2.

An easy but useful consequence of Theorem 1.2 is the following.

Corollary 1.3. *A finite irreducible chain is always recurrent.*

The proof is left as an exercise.

Problems

1. Consider a random walk on $\{0, 1, \ldots, 5\}$. The walker flips a fair coin and goes to the right if it is tail and to the left if it is head. If the walker hits the boundaries 0 or 5, then at the following step he goes to 1 or 4, respectively. Write the transition matrix for this random walk.

2. Assume that there are ten balls, five black and five white. The ten balls are distributed among two urns A and B with five balls each. At each step one ball is picked at random from each urn and they are switched from one urn to the other. Let X_n be the number of white balls at time n in urn A. Write the transition matrix for this chain.

3. A gambler has the following strategy. He bets 1\$ unless he has won the last two bets in which case he bets 2\$. Let X_n be the fortune of the gambler at time n. Is this a Markov chain?

4. At the roulette game there are 38 equally likely outcomes: numbers 0 through 36 and 00. These 38 possible outcomes are divided into 18 numbers that are red, 18 that are black, and 0 and 00 that are green. Assume that a gambler bets repeatedly on black, 1\$ each time. Let X_n be the fortune of the gambler after n bets. Assume also that the gambler quits when he is broke or when his fortune is 10\$. Write the transition probabilities for this chain.

5. Consider r balls labeled from 1 to r. Some balls are in box 1 and some balls are in box 2. At each step a number is chosen at random and the corresponding ball is moved from its box to the other box. Let X_n be the number of balls in box 1 after n steps. The chain X_n is called the Ehrenfest chain. Write the transition probabilities for this chain.

6. Assume that i and j are in the same recurrent class.
   Show that

$$\sum_{n=1}^{\infty} p_n(i, j) = +\infty.$$

7. Prove that if the distribution of X_0 and the $p(i, j)$ are given then the distribution of the Markov chain is completely determined. More precisely, prove that for any $n \geq 0$ and any $i_0, i_1, \ldots, i_n$ in $\mathcal{S}$ we have

$$P(X_0 = i_0, X_1 = i_1, \ldots, X_n = i_n) = P(X_0 = i_0)p(i_0, i_1)p(i_1, i_2) \ldots p(i_{n-1}, i_n).$$

8. Prove that the relation "to be in the same class as" is an equivalence relation on $\mathcal{S}$.

9. Consider a Markov chain with an absorbing state 0. That is, $p(0, 0) = 1$. Assume also that every state i we have $p(i, 0) > 0$.

(a) Is this be an irreducible chain?

(b) Show that all states except 0 are transient for this chain.

10. Consider the Markov chain on $\{1, 2, 3, 4, 5, 6\}$ having transition matrix

$$
\begin{pmatrix}
\frac{1}{6} & \frac{5}{6} & 0 & 0 & 0 & 0 \\
\frac{1}{3} & \frac{2}{3} & 0 & 0 & 0 & 0 \\
0 & 0 & \frac{1}{2} & 0 & \frac{1}{2} & 0 \\
\frac{1}{4} & \frac{1}{4} & 0 & 0 & \frac{1}{4} & \frac{1}{4} \\
0 & 0 & \frac{1}{2} & 0 & \frac{1}{2} & 0 \\
0 & \frac{1}{6} & 0 & \frac{1}{6} & \frac{1}{6} & \frac{1}{2}
\end{pmatrix}
$$

Find all the classes and determine which are transient and which are recurrent.

11. Consider a Markov chain on $\{0, 1, 2, \ldots\}$ for which 0 is a trap, that is, $p(0,0) = 1$. Assume also that $\{1, 2, \ldots\}$ is another class and that $p(i, 0) > 0$ for some $i \geq 1$.

(a) What can you say about the recurrence of each class.

(b) Can you guess the possible evolutions of such a chain?

12. Consider the Markov chain on $\{0, 1, 2, 3\}$ having transition matrix

$$
\begin{pmatrix}
0 & \frac{1}{2} & \frac{1}{2} & 0 \\
0 & 0 & 0 & 1 \\
0 & 1 & 0 & 0 \\
0 & 0 & \frac{1}{2} & \frac{1}{2}
\end{pmatrix}
$$

Find all the classes and determine which are transient and which are recurrent.

13. Give an example of a closed infinite class which is transient.

14. Show that the Ehrenfest chain defined in Problem 5 is irreducible and recurrent.

15. Consider the one dimensional random walk. Prove that if $p = 0$ or 1 then all states are transient.

16. Show that if $p \in (0, 1)$ then the one dimensional random walk is irreducible.

2 Birth and Death Chains

A birth and death chain is a Markov chain on the positive integers $S = \{0, 1, \ldots\}$. The transition probabilities are

$$
p(i, i+1) = p_i \qquad p(i, i) = r_i \qquad p(i, i-1) = q_i.
$$

where $p_i + q_i + r_i = 1$ for each $i \in S$. We assume that $q_i > 0$ for $i \geq 1$, $p_i > 0$ for all $i \geq 0$ and $r_i \geq 0$ for all $i \geq 0$.

Observe that in one time step the chain may move one unit to the left or to the right or stay put.

Next we compute the probability for the chain to reach state i before state j. Note that this is a generalization of the probability of ruin. Recall that F_k is the time of first visit to state k.

Proposition 2.1. *Let $0 \leq i < m < j$, the probability that starting at m the birth and death chain visits i before j is*

$$P(F_i < F_j | X_0 = m) = \frac{\sum_{k=m}^{j-1} P_k}{\sum_{\ell=i}^{j-1} P_\ell},$$

where $P_0 = 1$ and for $\ell \geq 1$

$$P_\ell = \frac{q_1}{p_1} \frac{q_2}{p_2} \dots \frac{q_\ell}{p_\ell}.$$

Note that Proposition 2.1 is a generalization of the ruin problem.

Proof of Proposition 2.1. We introduce the following notation. Let $i < k < j$ and

$$u(k) = P(F_i < F_j | X_0 = k).$$

Set $u(i) = 1$ and $u(j) = 0$. We condition on the first step to get

$$u(k) = P(F_i < F_j; X_1 = k - 1 | X_0 = k) + P(F_i < F_j; X_1 = k | X_0 = k) +$$
$$P(F_i < F_j; X_1 = k + 1 | X_0 = k).$$

By the Markov property this implies that

$$u(k) = q_k u(k-1) + r_k u(k) + p_k u(k+1) \text{ for all } k \in (i, j)$$

Replacing $u(k)$ by $(q_k + r_k + p_k)u(k)$ on the left-hand side we get

$$u(k+1) - u(k) = \frac{q_k}{p_k}(u(k) - u(k-1)).$$

We iterate the preceding equality to get

$$u(k+1) - u(k) = \frac{q_k}{p_k} \frac{q_{k-1}}{p_{k-1}} \dots \frac{q_{i+1}}{p_{i+1}}(u(i+1) - u(i)).$$

It is now convenient to introduce the following notation

$$P_0 = 1$$

$$P_k = \Pi_{l=1}^{k} \frac{q_l}{p_l} \text{ for all } k \geq 1.$$

Note that

$$\frac{q_k}{p_k} \frac{q_{k-1}}{p_{k-1}} \cdots \frac{q_{i+1}}{p_{i+1}} = \frac{P_k}{P_i}.$$

Hence,

$$u(k+1) - u(k) = \frac{P_k}{P_i}(u(i+1) - u(i)), \tag{2.1}$$

for $i < k < j$. Observe that (2.1) holds for $k = i$ as well. Summing (2.1) for all k in $[i, j)$ we get

$$\sum_{k=i}^{j-1} (u(k+1) - u(k)) = (u(i+1) - u(i)) \sum_{k=i}^{j-1} \frac{P_k}{P_i} \tag{2.2}$$

Note now that all the terms in the l.h.s. sum cancel except for the first and last. That is,

$$\sum_{k=i}^{j-1} (u(k+1) - u(k)) = u(j) - u(i).$$

Using the last equality and $u(i) = 1$, $u(j) = 0$ in (2.2) yields

$$-1 = (u(i+1) - u(i)) \sum_{k=i}^{j-1} \frac{P_k}{P_i}.$$

Hence,

$$u(i) - u(i+1) = \frac{1}{\sum_{k=i}^{j-1} \frac{P_k}{P_i}} = \frac{P_i}{\sum_{\ell=i}^{j-1} P_\ell}.$$

So (2.1) can be written as

$$u(k) - u(k+1) = \frac{P_k}{\sum_{\ell=i}^{j-1} P_\ell}.$$

Summing the preceding equality on all $k = m, \ldots, j - 1$ we get

$$\sum_{k=m}^{j-1} u(k) - u(k+1) = u(m) - u(j) = \frac{\sum_{k=m}^{j-1} P_k}{\sum_{\ell=i}^{j-1} P_\ell}.$$

Since $u(j) = 0$ we have

$$u(m) = P(F_i < F_j | X_0 = m) = \frac{\sum_{k=m}^{j-1} P_k}{\sum_{\ell=i}^{j-1} P_\ell}.$$

This completes the proof of Proposition 2.1.

We now use Proposition 2.1 to get a convenient recurrence criterion for birth and death chains. Recall that $q_i > 0$ for $i \geq 1$, $p_i > 0$ for all $i \geq 0$. We make the additional assumption that $q_0 = 0$ so that we have a reflective boundary at the origin. Under these assumptions the birth and death chain is irreducible, see the problems.

Proposition 2.2. *A birth and death chain on the positive integers with a reflective boundary at the origin is recurrent if and only if*

$$\sum_{k=1}^{\infty} P_k = \infty.$$

Proof of Proposition 2.2. Since the chain is irreducible it is enough to show that state 0 is recurrent.

By Proposition 2.1 we have

$$P(F_0 < F_n | X_0 = 1) = \frac{\sum_{k=1}^{n-1} P_k}{\sum_{\ell=0}^{n-1} P_\ell} = \frac{\sum_{k=0}^{n-1} P_k - P_0}{\sum_{\ell=0}^{n-1} P_\ell} = 1 - \frac{P_0}{\sum_{\ell=0}^{n-1} P_\ell} \qquad (2.3)$$

for $n \geq 2$. Note that $S_n = \sum_{\ell=0}^{n-1} P_\ell$ is an increasing sequence. It either converges to a finite number or goes to infinity. In any case

$$\lim_{n \to \infty} 1 - \frac{P_0}{\sum_{\ell=0}^{n-1} P_\ell} = 1 - \frac{P_0}{\sum_{\ell=0}^{\infty} P_\ell}$$

where $\frac{P_0}{+\infty}$ is taken to be 0.

Hence, the l.h.s. of (2.3) $P(F_0 < F_n | X_0 = 1)$ must also have a limit as n goes to infinity. This limit can be written as a probability as we now show. Since the chain moves one unit at most at a time we must have $F_2 < F_3 < \ldots$. That is, $(F_n)_{n \geq 1}$ is a strictly increasing sequence of integers. Therefore this sequence must go to infinity as n goes to infinity. It turns out that this is enough to show that

$$\lim_{n \to \infty} P(F_0 < F_n | X_0 = 1) = P(F_0 < +\infty | X_0 = 1).$$

We will not give a formal proof of this fact. Letting n go to infinity in (2.3) yields

$$P(F_0 < \infty | X_0 = 1) = 1 - \frac{P_0}{\sum_{\ell=0}^{\infty} P_\ell}.$$

So $P(F_0 < \infty | X_0 = 1) = 1$ if and only if $\sum_{\ell=0}^{\infty} P_\ell = \infty$. To conclude the proof we will show that $P(F_0 < \infty | X_0 = 1) = 1$ if and only if $P(F_0 < \infty | X_0 = 0) = 1$.
We have

$$P(F_0 < \infty | X_0 = 0) = r_0 + p_0 P(F_0 < \infty | X_0 = 1).$$

To see the preceding equality, observe that at time 1 either the chain stays at 0 with probability r_0 and then $F_0 = 1$ or the chain jumps to 1 with probability p_0 and by the Markov property we may consider that 1 is the initial state. Since $r_0 + p_0 = 1$, $P(F_0 < \infty | X_0 = 0) = 1$ if and only if $P(F_0 < \infty | X_0 = 1) = 1$ (why?). We have already shown that this in turn is equivalent to $\sum_{\ell=0}^{\infty} P_\ell = \infty$. This completes the proof of Proposition 2.2.

We now give an application of Proposition 2.2.

Example 2.1. Consider the random walk on the positive integers. That is, consider the birth and death chain with $p_i = p$ for $i \geq 1$ and $q_i = q$ for $i \geq 1$, $p_0 = 1$ and $q_0 = 0$. For what values of p is the random walk on the half line recurrent?
We have $P_k = (\frac{q}{p})^k$ and so $\sum_{k \geq 1} (\frac{q}{p})^k$ is a geometric series with ratio $\frac{q}{p}$. Thus, this series converges if and only if $\frac{q}{p} < 1$. Since $q = 1 - p$, we conclude that the random walk on the half line is recurrent if and only $p \leq \frac{1}{2}$.

We next show that we can generalize the preceding result.

Corollary 2.1. *Consider a birth and death chain with reflecting boundary at 0 (i.e., $q_0 = 0$). Assume that*

$$\lim_{n \to \infty} \frac{q_n}{p_n} = \ell.$$

- *If $\ell < 1$, the chain is transient.*
- *If $\ell > 1$, the chain is recurrent.*

Proof of Corollary 2.1. By Proposition 2.2 the chain is recurrent if and only if the series $\sum_{k=1}^{\infty} P_k$ diverges. Note that

$$\frac{P_{k+1}}{P_k} = \frac{q_{k+1}}{p_{k+1}}.$$

We assume that the r.h.s. has a limit ℓ. By the ratio test the series $\sum_{k=1}^{\infty} P_k$ converges for $\ell < 1$ and diverges for $\ell > 1$. The test is not conclusive for $\ell = 1$. This completes the proof of Corollary 2.1.

2.1 The Coupling Technique

Using Corollary 2.1 it is easy to see that if $\lim_{n\to\infty} p_n > \frac{1}{2}$ then the chain is transient. If $\lim_{n\to\infty} p_n < \frac{1}{2}$, the chain is recurrent. If the limit is equal to $\frac{1}{2}$, the chain can be recurrent or transient, see the problems. However, we can prove the following.

Example 2.2. Assume that there is a positive integer N such that for all $n \geq N$

$$p_n \leq \frac{1}{2}.$$

Then the chain is recurrent.

Recurrence will be the proved using the so-called *coupling technique*. We construct the chain $(X_n)_{n\geq 0}$ with birth probabilities p_n and the symmetric chain $(X'_n)_{n\geq 0}$ on the same probability space. To simplify matters we will take $r_n = 0$ for all $n \geq 0$ but the same ideas apply in general.

Let $(U_n)_{n\geq 1}$ be a sequence of independent uniform random variables on $(0, 1)$. That is, for $n \geq 1$ and $x \in (0, 1)$

$$P(U_n \leq x) = x.$$

Let $X_0 = X'_0 = N$. There are two cases to consider.

- If $X_1 = N - 1$, then the chain will return to N with probability 1. This is so because the chain is trapped in the finite set $\{0, 1, \dots, N\}$ and is irreducible. Therefore, it is recurrent.
- If $X_1 = N + 1$, then we use the coupling to show that the chain returns to N with probability 1. Assume $X_n = i \geq N + 1$ for $n \geq 1$. If $U_n \leq p_i$, we set $X_{n+1} = X_n + 1$. If $U_n > p_i$, we set $X_{n+1} = X_n - 1$. Observe that this gives the right probabilities for the jumps to the left and to the right: $p(i, i + 1) = p_i = P(U_n \leq p_i)$.

We use the *same* U_n for X'_n: If $X'_n = i$ and $U_n \leq \frac{1}{2}$, then $X'_{n+1} = X'_n + 1$ while if $U_n > \frac{1}{2}$, then $X'_{n+1} = X'_n - 1$.

Here are two important remarks.

1. Since $p_i \leq \frac{1}{2}$ if $U_n \leq p_i$ then $U_n \leq \frac{1}{2}$. That is, if $X_n = X'_n = i$ and X_n jumps to the right so does X'_n.

2. We start with $X_0 = X_0'$ and $X_n = X_n'$ until X_n' jumps to the right and X_n jumps to the left. This happens if $p_n < U_n < \frac{1}{2}$. The important point is that if $X_0 = X_0'$ then $X_n' - X_n \geq 0$ is always even or zero. For the chains either move in the same direction and the difference $X_n' - X_n$ does not change or they move in opposite directions and that adds $+2$ or -2 to the difference. In particular in order to have $X_n > X_n'$ for some n the two chains would have to meet at some prior time. But when the chains meet we are back to step 1 above. Hence, at all times $n \geq 0$ we have $X_n \leq X_n'$.

The final step is to note that $(X_n')_{n \geq 0}$ is a recurrent chain and therefore returns to N with probability 1. Since $X_n \leq X_n'$ for all $n \geq 1$ the chain $(X_n')_{n \geq 0}$ returns to N as well. This completes the proof.

Example 2.3. By the preceding example if there is J such that $p_i \leq \frac{1}{2}$ for all $i > J$ the chain is recurrent. In this example we show that a chain may have $p_i > \frac{1}{2}$ for every i and still be recurrent. Let

$$p_j = \frac{1}{2} + s(j), \quad q_j = \frac{1}{2} - s(j), \quad s(j) \geq 0 \text{ for all } j \geq 0, \text{ and } \lim_{j \to \infty} s(j) = 0.$$

That is, we assume that p_j approaches $\frac{1}{2}$ from above. This example will show that if p_j approaches $\frac{1}{2}$ fast enough then the chain is recurrent. More precisely, we will show that if the series $\sum_{j=0}^{\infty} s(j)$ converges then the chain is recurrent.

By Proposition 2.2 a birth and death chain is recurrent if and only if $\sum_{k \geq 0} P_k = \infty$, where

$$P_k = \Pi_{l=1}^{k} \frac{q_l}{p_l} \text{ for all } k \geq 1.$$

We have

$$\frac{q_j}{p_j} = \frac{\frac{1}{2} - s(j)}{\frac{1}{2} + s(j)} = 1 - \frac{2s(j)}{\frac{1}{2} + s(j)}.$$

We will need the following fact about infinite products. Consider a sequence s_j in $(0,1)$, we have that

$$\Pi_{j=0}^{\infty}(1 - s_j) > 0 \text{ if and only if } \sum_{j=0}^{\infty} s_j < \infty.$$

For a proof see, for instance, the Appendix. So

$$\lim_{k \to \infty} P_k > 0 \text{ if and only if } \sum_{j=0}^{\infty} \frac{2s(j)}{\frac{1}{2} + s(j)} < \infty.$$

Hence, if $\sum_{j=0}^{\infty} \frac{2s(j)}{\frac{1}{2}+s(j)}$ converges, then $\lim_{k\to\infty} P_k > 0$. Therefore, the series $\sum_{k\geq 0} P_k$ diverges (why?). Recall that $\lim_{j\to\infty} s(j) = 0$, hence as j goes to infinity

$$\frac{2s(j)}{\frac{1}{2}+s(j)} \sim 4s(j).$$

Therefore, $\sum_{j=0}^{\infty} \frac{2s(j)}{\frac{1}{2}+s(j)}$ converges if and only if $\sum_{j=0}^{\infty} s(j)$ converges. In particular, if

$$s(j) \sim \frac{C}{j^\alpha}$$

for some positive C and $\alpha > 1$ the birth and death chain is recurrent.

2.2   An Application: Quorum Sensing

"Quorum sensing" describes a strategy used by some bacteria under which the bacteria multiply until a critical mass is reached. At that point the bacteria turn on their virulence genes and launch an attack on their host. Several human diseases (such as cholera) are caused by quorum sensing bacteria. We are interested in the following question. Is it obvious that there is always strength in numbers? We exhibit a simple probability model that shows that things may be less simple than they appear.

We now describe the model. We start with one individual. There are two phases in the process. In the first phase either the population (of bacteria, for instance) disappears or it gets to the quorum N where N is a natural number. We will use a birth and death chain to model this first phase. If the population reaches N, then the second phase kicks in. In the second phase we assume that each one of the N individuals has a probability ρ of being successful. Success may mean different things in different situations. For a bacterium it may mean not being eliminated by the host. There is evidence that in the case of cholera not only do the bacteria multiply but they also evolve into more pathogenic strains before launching their attack on the host. So we may think of N as the number of strains rather than the number of individual bacteria and we may think of ρ as the probability that a given strain escapes the immune system of the host.

Let A_N be the event that the population or the number of strains eventually reaches N (starting with one individual). The alternative to the event A_N is that the population gets killed off before reaching N. In order for at least one individual in the population to be successful we first need the population to reach N and then we need at least one of the N individuals to be successful. Note that the probability that all N individuals are (independently) unsuccessful is $(1 - \rho)^N$. Hence, the probability that at least one of the N individuals is successful is $1 - (1 - \rho)^N$.

Therefore, the probability that the population will reach the quorum N and then have a successful individual is

$$f(N, \rho) = P(A_N)(1 - (1 - \rho)^N),$$

where we are assuming that the two phases are independent. The function f represents the probability of success under a quorum sensing strategy, N represents the quorum. We are interested in f as a function of N. If there is always "strength in numbers," then f must be always increasing. In fact, we will show that f may be increasing, decreasing or neither.

We use a birth and death chain to model the first phase. Assume that there are n individuals at some point where $1 \le n < N$. Then there is a death with probability q_n or a birth with probability p_n. We also assume that if the chain gets to 0 before getting to N then it stays there. By Proposition 2.1 we have

$$P(A_N) = \frac{1}{\sum_{i=0}^{N-1} P_i},$$

where $P_0 = 1$ and for $i \ge 1$

$$P_i = \Pi_{k=1}^i \frac{q_k}{p_k}.$$

We now assume that the birth and death probabilities are constant. That is, $p_n = p$ for all n in $[1, N - 1]$ and therefore $q_n = 1 - p = q$ for all such n. Let $r = q/p$. We get $P_i = r^i$,

$$P(A_N) = \frac{r - 1}{r^N - 1}$$

and

$$f(N, \rho) = P(A_N)(1 - (1 - \rho)^N) = \frac{r - 1}{r^N - 1}(1 - (1 - \rho)^N).$$

A factorization yields

$$f(N, \rho) = \rho \frac{1 + (1 - \rho) + \ldots + (1 - \rho)^{N-1}}{1 + r + \ldots + r^{N-1}}.$$

Using this last expression and some simple algebra it is not difficult to show that f is decreasing as a function of N if $r > 1 - \rho$ and increasing if $r < 1 - \rho$, see the problems. Note that if $r \ge 1$ then f is always decreasing. Hence, there is a dramatic change depending whether ρ is smaller than or larger than $1 - r$ provided $r < 1$.

In fact, in the problems section we will give an example of another birth and death chain for which the function f is neither increasing nor decreasing. So to the question "Is there strength in numbers?" the answer is "Not always!"

Problems

1. Decide whether the following chains are recurrent or transient:

(a) $p_i = \frac{100}{100+i}$ and $q_i = \frac{i}{100+i}$ for all $i \geq 0$

(b) $p_i = \frac{3i+1}{4i+1}$ and $q_i = \frac{i}{4i+1}$ for $i \geq 1$.

2. Consider the Ehrenfest chain. It is defined as follows. There are r balls labeled from 1 to r. Some balls are in box 1 and some balls are in box 2. At each step a number is chosen at random and the corresponding ball is moved from its box to the other box. Let X_n be the number of balls in box 1 after n steps. Show that the Ehrenfest chain is a birth and death chain.

3. Are the following birth and death chains recurrent or transient?

(a) $p_i = \frac{100}{100+i}$ and $q_i = \frac{i}{100+i}$ for all $i \geq 0$.

(b) $p_i = \frac{3i+1}{4i+1}$ and $q_i = \frac{i}{4i+1}$ for $i \geq 1$.

4. Assume that

$$\lim_{n \to \infty} p_n = p.$$

(a) Show that if $p < \frac{1}{2}$ the chain is recurrent. (Use Corollary 2.1.)

(b) Show that if $p > \frac{1}{2}$ the chain is transient.

5. The following two examples show that if $\lim_{i \to \infty} p_i = \frac{1}{2}$ the chain may be transient or recurrent.

(a) Let

$$p_i = \frac{i+2}{2(i+1)} \qquad q_i = \frac{i}{2(i+1)}.$$

Is this chain transient or recurrent?

(b) Let

$$p_i = \frac{i+1}{2i+1} \qquad q_i = \frac{i}{2i+1}.$$

Is this chain transient or recurrent?

6. Consider a birth and death chain with J such that $p_i \geq \frac{5}{8}$ for all $i \geq J$. Show that this chain is transient. Use the coupling method.

7. Consider a birth and death chain with J such that $p_i \geq \frac{1}{2} + a$ where a is a fixed number in $(0, \frac{1}{2})$.

(a) Show that the chain is transient. Use the coupling method.
(b) What can you say about the chain when $a = 0$?

8. From Example 2.3 we know that if

$$p_j = \frac{1}{2} + s(j) \text{ and } s(j) \sim \frac{C}{j^\alpha}$$

for some $\alpha > 1$ the chain is recurrent.
   Use this criterion to show that the chain with

$$p(j) = \frac{j^2 + 2}{2(j^2 + 1)}$$

is recurrent.

9. Let f and g be increasing functions. Let X be a random variable. We want to prove that $f(X)$ and $g(X)$ are positively correlated. That is we want to show that

$$E(f(X)g(X)) \geq E(f(X))E(g(X)).$$

We do a proof by coupling. Define X and Y to be two independent copies of the same random variable.

(a) Show that for all x and y we have

$$(f(x) - f(y))(g(x) - g(y)) \geq 0.$$

(b) Show that

$$E((f(X) - f(Y))(g(X) - g(Y))) \geq 0.$$

(c) Expanding (b) show that $f(X)$ and $g(X)$ are positively correlated.

10. Consider a birth and death chain with absorbing barriers at 0 and 3 with $p_1 = \frac{1}{3}$, $q_1 = 2/3$, $p_2 = 3/4$, $q_2 = \frac{1}{4}$. Write a relation between D_i, D_{i-1} and D_{i+1} (see the ruin problem). Compute D_1 and D_2.

11. Consider a birth and death chain on $[0, N]$, where N is a positive integer, with absorbing barriers at 0 and N. That is, $p(0,0) = 1$ and $p(N, N) = 1$. We also assume that $p_i > 0$ and $q_i > 0$ for all i in $(0, N)$.

(a) Show that this chain has two recurrent classes and one transient class.

(b) Let D_i be the (random) time it takes for the chain starting at i in $[0, N]$ to get absorbed at 0 or N. Show that if $\{D_i = \infty\}$ then the chain must visit at least one state in $(0, N)$ infinitely often.

(c) Use (b) to show that with probability one D_i is finite.

12. Consider the quorum sensing model with constant $p_n = p = 0.55$.

(a) Graph the function $f(N, \rho)$ for $\rho = 0.17$.
(b) Graph the function $f(N, \rho)$ for $\rho = 0.19$.
(c) Interpret (a) and (b).

13. Consider the quorum sensing model for which

$$p_n = \frac{n+1}{2n+1} \text{ and } q_n = \frac{n}{2n+1},$$

for n in $[1, N - 1]$.

(a) Show that for $i \geq 0$, $P_i = \frac{1}{i+1}$.
(b) Show that

$$P(A_N) = \frac{1}{\sum_{i=1}^{N} \frac{1}{i}},$$

and therefore

$$f(N, \rho) = \frac{1}{\sum_{i=1}^{N} \frac{1}{i}} (1 - (1 - \rho)^N).$$

(c) Show by graphing f as a function of N for a fixed ρ that f is increasing for small values of ρ, decreasing for large values of ρ and neither for intermediate values of ρ.

14. (a) Assume that a, b, c, d are strictly positive numbers. Show that if $c < d$ then

$$\frac{a}{b} > \frac{a+c}{b+d}.$$

(b) Use (a) to show that

$$f(N) = \rho \frac{1 + (1 - \rho) + \ldots + (1 - \rho)^{N-1}}{1 + r + \ldots + r^{N-1}}$$

is increasing (as a function of N) if $r < 1 - \rho$ and decreasing for $r > 1 - \rho$.

Notes

There are many good books on Markov chains. At an elementary level see, for instance, Feller (1968), Hoel et al. (1972) or Karlin and Taylor (1975). A more advanced book is Bhattacharya and Waymire (1990).

References

Bhattacharya, R., Waymire, E.: Stochastic Processes with Applications. Wiley, New York (1990)
Feller, W.: An Introduction to Probability Theory and its Applications, vol. I, 3rd edn. Wiley, New york (1968)
Hoel, P., Port, S., Stone, C.: Introduction to Stochastic Processes. Houghton Mifflin, Boston (1972)
Karlin, S., Taylor, H.: A First Course in Stochastic Processes, 2nd edn. Academic, New York (1975)
Port, S.C.: Theoretical Probability for applications. Wiley (1994)

Chapter 6
Stationary Distributions for Discrete Time Markov Chains

We continue the study of Markov chains initiated in Chap. 5. A stationary distribution is a stochastic equilibrium for the chain. We find conditions under which such a distribution exists. We are also interested in conditions for convergence to a stationary distribution.

1 Convergence to a Stationary Distribution

1.1 Convergence and Positive Recurrence

We will use the notation and results from Sect. 1.1 of the preceding chapter. Consider a discrete time Markov chain $(X_n)_{n\geq0}$. Let i and j be two states. As n goes to infinity, when does

$$p_n(i, j) = P(X_n = j | X_0 = i)$$

converge? In words, does the distribution of X_n converge as n goes to infinity?
    It is easy to think of examples for which the answer is no.

Example 1.1. Consider the Markov chain on $\{0, 1\}$ with transition probabilities

$$\begin{pmatrix} 0 & 1 \\ 1 & 0 \end{pmatrix}$$

Note that $p_n(0,0) = 0$ if n is odd and $p_n(0,0) = 1$ if n is even (why?). Thus, $p_n(0,0)$ does not converge.

© Springer Science+Business Media New York 2014
R.B. Schinazi, *Classical and Spatial Stochastic Processes: With Applications to Biology*, DOI 10.1007/978-1-4939-1869-0_6

The preceding example illustrates the importance of the notion of periodicity that we now introduce.

Definition 1.1. Assume that j is a recurrent state for the chain X_n with transition probabilities $p(i, j)$. Let

$$I_j = \{n \geq 1 : p_n(j, j) > 0\}$$

and let d_j be the greatest common divisor (g.c.d.) of I_j. We call d_j the period of state j.

Note that if j is recurrent we know that $\sum_{n \geq 1} p_n(j, j) = \infty$. In particular I_j is not empty and d_j is well defined. In Example 1.1 we have the periods $d_0 = d_1 = 2$ (why?). We now show that periodicity is a class property.

Proposition 1.1. *All states in the same recurrent class have the same period.*

Proof of Proposition 1.1. Assume i and j are in the same recurrent class. Then there are two integers n and m such that

$$p_n(i, j) > 0 \qquad p_m(j, i) > 0.$$

We have

$$p_{m+n}(j, j) \geq p_m(j, i) p_n(i, j) > 0$$

so $m + n$ belongs to I_j and d_j must divide $m + n$. Let n_0 be in I_i. Then

$$p_{m+n+n_0}(j, j) \geq p_m(j, i) p_{n_0}(i, i) p_n(i, j) > 0$$

so $m + n + n_0$ is in I_j and therefore d_j divides $m + n + n_0$. Since d_j divides $m + n$ it must also divide n_0. Hence, d_j divides every element in I_i. It must divide d_i. But i and j play symmetric roles so d_i must divide d_j. Therefore, $d_i = d_j$. This completes the proof of Proposition 1.1.

Definition 1.2. A class with period 1 is said to be aperiodic.

Recall that given $X_0 = i$, $f_n(i, i)$ is the probability that the chain returns for the first time to i at time $n \geq 1$. A state i is said to be recurrent if given $X_0 = i$ the chain eventually returns to i with probability 1. This is equivalent to

$$\sum_{n \geq 1} f_n(i, i) = 1.$$

A state i is recurrent if and only if (see Theorem 1.1 in the preceding chapter)

$$\sum_{n \geq 1} p_n(i, i) = +\infty.$$

Let F_i be the (random) time of the first visit to state i. Hence, for $n \geq 1$

$$P(F_i = n | X_0 = i) = f_n(i, i).$$

Therefore, the expected value of F_i is

$$E(F_i | X_0 = i) = \sum_{n \geq 1} n f_n(i, i)$$

where the series is possibly infinite.

Definition 1.3. Let i be a recurrent state. If $E(F_i | X_0 = i) < +\infty$, then i is said to be positive recurrent. If $E(F_i | X_0 = i) = +\infty$, then i is said to be null recurrent.

Example 1.2. We have shown in a previous chapter that the simple symmetric random walk on $\mathbf{Z}$ is null recurrent. On the other hand, we will show in the sequel that a finite irreducible Markov chain is always positive recurrent.

We now state the main convergence result for Markov chains.

Theorem 1.1. *Consider an irreducible aperiodic recurrent Markov chain. Then, for all i and j*

$$\lim_{n \to \infty} p_n(i, j) = \frac{1}{E(F_j | X_0 = j)}.$$

If state j is null recurrent we have $E(F_j | X_0 = j) = +\infty$. In this case we set

$$\frac{1}{E(F_j | X_0 = j)} = 0.$$

Note that there are several hypotheses for the theorem to hold. Note also that the limit does not depend on the initial state i. For a proof of Theorem 1.1 see Sect. 3.1 in Karlin and Taylor (1975). We now state several interesting consequences of Theorem 1.1.

Corollary 1.1. *All states in a recurrent class are either all positive recurrent or all null recurrent.*

Proof of Corollary 1.1. We will give this proof in the particular case when the class is aperiodic. For the general case see Sect. 3.1 in Karlin and Taylor (1975).

Assume that i and j belong to the same aperiodic recurrent class. Hence, there are natural numbers n and m such that

$$p_n(i, j) > 0 \text{ and } p_m(j, i) > 0.$$

For any natural k

$$p_{n+k+m}(i, i) \geq p_n(i, j) p_k(j, j) p_m(j, i).$$

Let k go to infinity in the inequality above and use Theorem 1.1 to get

$$\frac{1}{E(F_i|X_0 = i)} \geq p_n(i, j)\frac{1}{E(F_j|X_0 = j)}p_m(j, i).$$

Assume now that j is positive recurrent. The r.h.s. is strictly positive and so i is also positive recurrent. On the other hand, if i is null recurrent, then the l.h.s. is 0 and therefore j is also null recurrent. This concludes the proof of Corollary 1.1.

Corollary 1.2. *If state j is null recurrent or transient, then for all i*

$$\lim_{n\to\infty} p_n(i, j) = 0.$$

Proof of Corollary 1.2. Assume that j is null recurrent. If j is aperiodic by Theorem 1.1

$$\lim_{n\to\infty} p_n(i, j) = 0.$$

If j is null recurrent and periodic, the result still holds but the argument is more involved and we omit it.

If j is transient, then Corollary 1.2 of the preceding chapter states that for all i

$$\sum_{n=1}^{\infty} p_n(i, j) < +\infty.$$

Hence, $\lim_{n\to\infty} p_n(i, j) = 0$ and this completes the proof of Corollary 1.2.

Example 1.3. Consider a random walk on **Z**. It moves one step to the right with probability p and one step to the left with probability $q = 1 - p$. We know that the random walk is transient for $p \neq q$ and null recurrent for $p = q = \frac{1}{2}$. We now check that Corollary 1.2 holds in this example.

It is easy to see that

$$p_{2n}(0, 0) = \binom{2n}{n} p^n q^n.$$

By Stirling's formula we get as n goes to infinity that

$$p_{2n}(0, 0) \sim \frac{(4pq)^n}{\sqrt{\pi n}}.$$

Since $4pq \leq 1$ for all p in $[0, 1]$ we have

$$\lim_{n\to\infty} p_{2n}(0, 0) = 0.$$

1.2   Stationary Distributions

We will show that for a positive recurrent chain the limits of $p_n(i, j)$ play an important role for the chain. We start with a definition.

Definition 1.4. Assume that X_n is a Markov chain $p(i, j)$. A probability distribution π is said to be a stationary distribution for X_n if for all j we have

$$\sum_i \pi(i)p(i, j) = \pi(j)$$

where the sum is taken over all states i.

Example 1.4. Consider a Markov chain with states in $\{0, 1\}$. Let the transition probabilities be

$$P = \begin{pmatrix} 1/2 & 1/2 \\ 1/3 & 2/3 \end{pmatrix}$$

Then π is a stationary distribution if

$$\pi(0)/2 + \pi(1)/3 = \pi(0)$$
$$\pi(0)/2 + \pi(1)2/3 = \pi(1).$$

Observe that this system may also be written with the following convenient matrix notation. Let π be the row vector $\pi = (\pi(0), \pi(1))$. The preceding system is equivalent to

$$\pi P = \pi.$$

Note that $\pi(0) = 2/3\pi(1)$. This together with $\pi(0) + \pi(1) = 1$ gives $\pi(0) = 2/5$ and $\pi(1) = 3/5$.

Next we are going to show that the distribution $\pi = (2/5, 3/5)$ is an equilibrium for this chain in the following sense. If X_0 has distribution π, i.e. $P(X_0 = 0) = 2/5$ and $P(X_0 = 1) = 3/5$, then at all times $n \geq 1$, X_n has also distribution π.

Proposition 1.2. *Assume that π is a stationary distribution for the Markov chain with transition probabilities $p(i, j)$. Then for any $n \geq 1$ and any state j*

$$\sum_i \pi(i)p_n(i, j) = \pi(j).$$

To prove Proposition 1.2 we will need the following.

Proposition 1.3. *For any* $n \geq 0$ *and states* i *and* j *we have*

$$p_{n+1}(i, j) = \sum_k p_n(i, k) p(k, j).$$

The proof of Proposition 1.3 follows from conditioning on $X_n = k$ for all states k and the Markov property. We can now turn to the proof of Proposition 1.2.

Proof of Proposition 1.2. We do an induction proof. By definition of stationarity we have that

$$\sum_i \pi(i) p_n(i, j) = \pi(j)$$

holds for $n = 1$. Assume the equality holds for n. By Proposition 1.3 we have

$$\sum_i \pi(i) p_{n+1}(i, j) = \sum_i \pi(i) \sum_k p_n(i, k) p(k, j).$$

Since the terms above are all positive we may change the order of summation to get

$$\sum_i \pi(i) p_{n+1}(i, j) = \sum_k \sum_i \pi(i) p_n(i, k) p(k, j).$$

By the induction hypothesis we have

$$\sum_i \pi(i) p_n(i, k) = \pi(k).$$

Hence,

$$\sum_i \pi(i) p_{n+1}(i, j) = \sum_k \pi(k) p(k, j) = \pi(j).$$

This completes the induction proof of Proposition 1.2.

Let X_0 be distributed according to π. That is, for any state i

$$P(X_0 = i | \pi) = \pi(i).$$

If π is stationary, then by the Markov property and Proposition 1.2

$$P(X_n = j | \pi) = \sum_i \pi(i) p_n(i, j) = \pi(j).$$

Therefore for any $n \geq 0$, X_n is distributed according to π. This is why such a distribution is called stationary. It is stationary in time.

Theorem 1.2. *Consider an irreducible positive recurrent Markov chain. Then, the chain has a unique stationary distribution π given by*

$$\pi(i) = \frac{1}{E(F_i | X_0 = i)}.$$

By Theorems 1.1 and 1.2 we see that for an irreducible aperiodic positive recurrent Markov chain $p_n(i, j)$ converges to $\pi(j)$ for all states i and j. Moreover, π is stationary. In other words, the chain converges towards the equilibrium π.

Proof of Theorem 1.2. We prove the theorem in the particular case when the number of states is finite and the chain is aperiodic. For the general case see Hoel et al. (1972).

Define

$$\pi(k) = \frac{1}{E(F_k | X_0 = k)}.$$

We are first going to show that π is a probability distribution. Since the chain is positive recurrent we know that $\pi(k) > 0$ for every k. We now show that the sum of the $\pi(k)$ is 1.

We have for all $n \geq 0$ and j

$$\sum_k p_n(j, k) = 1.$$

We let n go to infinity to get

$$\lim_{n \to \infty} \sum_k p_n(j, k) = 1.$$

Since we are assuming that there are finitely many states the sum is finite and we may interchange the sum and the limit. Hence,

$$\sum_k \lim_{n \to \infty} p_n(j, k) = 1.$$

By Theorem 1.1 we have

$$\sum_k \pi(k) = 1.$$

Hence, π is a probability distribution.

The second step is to show that π is stationary. By Proposition 1.3

$$p_{n+1}(i, j) = \sum_k p_n(i, k) p(k, j).$$

We let n go to infinity

$$\lim_{n\to\infty} p_{n+1}(i,j) = \lim_{n\to\infty} \sum_k p_n(i,k)p(k,j).$$

Since the sum is finite we may interchange the sum and the limit to get

$$\lim_{n\to\infty} p_{n+1}(i,j) = \sum_k \lim_{n\to\infty} p_n(i,k)p(k,j).$$

By Theorem 1.1 we have that

$$\pi(j) = \sum_k \pi(k)p(k,j).$$

This proves that π is stationary.

The last step is to prove that π is the unique stationary distribution. Assume that a is also a stationary distribution. By Proposition 1.2 we have

$$a(i) = \sum_j a(j)p_n(j,i).$$

We let n go to infinity and interchange the limit and the finite sum to get

$$a(i) = \sum_j a(j)\pi(i) = \pi(i)\sum_j a(j).$$

Since a is a probability distribution we have

$$a(i) = \pi(i)\sum_j a(j) = \pi(i).$$

Hence, a is necessarily equal to π. This completes the proof of Theorem 1.2.

Next we show that an irreducible chain that has a stationary distribution must be positive recurrent.

Theorem 1.3. *Consider an irreducible Markov chain. Assume that it has a stationary distribution. The chain must be positive recurrent.*

Proof of Theorem 1.3. Again we prove the theorem in the finite case but it holds in general.

By contradiction assume that the chain is transient or null recurrent. Then, by Corollary 1.2 we have for all i and j

$$\lim_{n\to\infty} p_n(i,j) = 0.$$

Let π be the stationary distribution of the chain. By Proposition 1.2 we have

$$\sum_i \pi(i) p_n(i, j) = \pi(j).$$

Letting n go to infinity and interchanging the finite sum and the limit we get

$$\pi(j) = \sum_i \pi(i) \lim_{n \to \infty} p_n(i, j) = 0,$$

for every j. But the sum of the $\pi(j)$ is 1. We have a contradiction. The chain must be positive recurrent.

Example 1.5. Consider a simple random walk on the integers. That is, $p(i, i+1) = p$ and $p(i, i-1) = 1 - p = q$. We know that the chain is transient if $p \neq q$ and is null recurrent if $p = q$. Since this is an irreducible chain Theorem 1.3 applies. For any p in $(0, 1)$ the random walk has no stationary distribution.

1.3 The Finite Case

In this section we consider a Markov chain X_n on a finite set. We start with an example.

Example 1.6. This example will show that if the chain is reducible (i.e., the chain has two or more classes) then we may have several stationary distributions.

Consider the Markov chain on $\{1, 2, 3, 4, 5\}$ with transition probabilities

$$P = \begin{pmatrix} 1/2 & 1/2 & 0 & 0 & 0 \\ 1/2 & 1/2 & 0 & 0 & 0 \\ 0 & 1/2 & 0 & 1/2 & 0 \\ 0 & 0 & 0 & 1/2 & 1/2 \\ 0 & 0 & 0 & 1/3 & 2/3 \end{pmatrix}$$

A stationary distribution π is a solution of

$$\pi P = \pi$$

where π is a row vector. Solving the system of equations gives $\pi(1) = \pi(2)$, $\pi(3) = 0$, $\pi(4) = 2\pi(5)/3$. Thus, there are infinitely solutions even with the restriction that the sum of the $\pi(i)$ is one. Note that there are two closed classes $C_1 = \{1, 2\}$ and $C_2 = \{4, 5\}$. Note that the chain may be restricted to C_1. Since C_1 is closed if the chain starts in C_1 it will stay there forever. The Markov chain restricted to C_1 has transition probabilities

$$\begin{pmatrix} 1/2 \ 1/2 \\ 1/2 \ 1/2 \end{pmatrix}$$

and there is a unique stationary distribution $\pi_1(1) = 1/2 \ \pi_1(2) = 1/2$. Moreover, this is an aperiodic chain (why?) and Theorem 1.1 holds. Thus,

$$\lim_{n \to \infty} p_n(i, j) = \pi_1(j) \text{ for } i, j \in C_1.$$

Likewise we may restrict the chain to $C_2 = \{4, 5\}$. This time the transition probabilities are

$$\begin{pmatrix} 1/2 \ 1/2 \\ 1/3 \ 2/3 \end{pmatrix}$$

We find the unique stationary distribution $\pi_2(4) = 2/5, \pi_2(5) = 3/5$. Again this is an aperiodic chain and therefore

$$\lim_{n \to \infty} p_n(i, j) = \pi_2(j) \text{ for } i, j \in C_2.$$

We extend π_1 and π_2 to the whole set by setting $\pi_1(3) = \pi_1(4) = \pi_1(5) = 0$ and $\pi_2(3) = \pi_2(1) = \pi_2(2) = 0$. It is easy that π_1 and π_2 are stationary distributions. Moreover, if we have two or more stationary distributions, then there are infinitely many stationary distributions. See the problems.

Observe also that $\pi(3)$ is zero for all the stationary distributions π. This must be so since state 3 is transient.

Theorem 1.4. *Assume that X_n is a finite Markov chain. Then X_n has at least one stationary distribution.*

For a proof of Theorem 1.4 see Levin et al. (2008).
We have the following consequences.

Corollary 1.1. *Assume that X_n is an irreducible finite Markov chain. Then all states are positive recurrent.*

Proof of Corollary 1.1. By Theorem 1.4 the chain has a stationary distribution. By Theorem 1.3 the chain must be positive recurrent. This proves Corollary 1.1.

Corollary 1.2. *Assume that X_n is a finite Markov chain. Then there are no null recurrent states.*

Proof of Corollary 1.2. Assume that i is a recurrent state. Then the class of i is closed. Therefore, if the chain starts in this class it will stay there at all times. The class must also be finite since the state space is finite. Hence, the restriction of the chain to this finite class is irreducible. Therefore, by Corollary 1.1 the chain restricted to this class is positive recurrent and so is state i. Hence, we have proved that if a state is recurrent it must be positive recurrent. This proves Corollary 1.2.

Note that Example 1.6 is the typical situation for a finite chain. Each closed class C has a stationary distribution π_C which can be extended to the whole space by setting $\pi_C(j) = 0$ for j not in C.

Problems

1. Consider a simple random walk on $\{0, \ldots, N\}$:

$$p(i, i+1) = p \quad p(i, i-1) = 1 - p \text{ for } 1 \le i \le N - 1, \quad p(0, 1) = 1 = p(N, N-1)$$

(a) Show that this chain is irreducible. What is the period?
(b) Is there a stationary distribution for this chain?
(c) Does $p_n(i, j)$ converge as n goes to infinity?

2. Assume that $p(0, 0) > 0$ and that this chain is irreducible. What is the period of this chain?

3. Consider a Markov chain with transition probabilities

$$\begin{pmatrix} 1/2 & 0 & 1/2 \\ 1/2 & 0 & 1/2 \\ 1/2 & 1/2 & 0 \end{pmatrix}$$

Show that for every i and j $p_n(i, j)$ converges as n goes to infinity.

4. Consider a random walk on circle marked with points $\{0, 1, 2, 3, 4\}$. The walker jumps one unit clockwise with probability $1/3$ or one unit counterclockwise with probability $2/3$. Find the proportion of time that the walker spends at 0.

5. Let X_n be the sum of n rolls of a fair die. Let Y_n be the integer rest of the division by 5 of X_n.

(a) Find the transition probabilities for the Markov chain Y_n.
(b) Find

$$\lim_{n \to \infty} P(X_n \text{ is a multiple of } 5).$$

6. Find the stationary distributions if the transition probabilities are

$$\begin{pmatrix} 1/3 & 2/3 & 0 \\ 1/2 & 0 & 1/2 \\ 1/6 & 1/3 & 1/2 \end{pmatrix}$$

7. Assume that a Markov chain has a trap at state 0. That is, $p(0, i) = 0$ for all $i \neq 0$. Show that the distribution π defined by $\pi(0) = 1$ and $\pi(i) = 0$ for all $i \neq 0$ is stationary.

8. (a) Assume that π_1 and π_2 are stationary distributions. Show that $t\pi_1 + (1-t)\pi_2$ is also stationary for all t in $[0,1]$.

    (b) Show that if a chain has two stationary distributions then it has infinitely many stationary distributions.

9. Let p be in $(0, 1)$. Consider a chain on the positive integers such that

$$p(i, i+1) = p \text{ and } p(i,0) = 1 - p \text{ for } i \geq 0$$

where p is a fixed number strictly between 0 and 1.

  (a) Prove that this chain is irreducible.

  (b) Find the periodicity.

  (c) Find a stationary distribution.

  (d) Prove that this is a positive recurrent chain.

10. The transition probabilities $p(i, j)$ are said to be doubly stochastic if

$$\sum_i p(i, j) = 1 \text{ and } \sum_j p(i, j) = 1.$$

Assume that there are N states. Find a stationary distribution.

11. Assume that a chain is irreducible aperiodic positive recurrent. Show that for every state j

$$\sum_i \frac{1}{E(F_i | X_0 = i)} p(i, j) = \frac{1}{E(F_j | X_0 = j)},$$

where F_k denotes the time of first visit to state k.

12. Assume that X_n is irreducible and recurrent. Show that for any i and j the random time the chain takes to go from i to j is finite with probability 1.

13. Consider the chain with transition probabilities

$$\begin{pmatrix} 0 & 1/2 & 1/2 & 0 \\ 0 & 1 & 0 & 0 \\ 0 & 0 & 1/3 & 2/3 \\ 0 & 0 & 1/2 & 1/2 \end{pmatrix}.$$

  (a) Find the stationary distributions of this chain.

  (b) Find all the limits of $p_n(i, j)$ that exist.

14. Same questions as above for

$$\begin{pmatrix} 1/2 & 1/2 & 0 & 0 \\ 1/5 & 4/5 & 0 & 0 \\ 1/2 & 0 & 0 & 1/2 \\ 0 & 0 & 1/2 & 1/2 \end{pmatrix}.$$

15. Consider a finite Markov chain on $\{0, 1\}$ with probability transitions

$$P = \begin{pmatrix} 1/3 & 2/3 \\ 1/2 & 1/2 \end{pmatrix}.$$

(a) Find the stationary distribution for this chain.

(b) Let M be

$$M = \begin{pmatrix} 1 & -4 \\ 1 & 3 \end{pmatrix}$$

and D be

$$D = \begin{pmatrix} 1 & 0 \\ 0 & -1/6 \end{pmatrix}.$$

Show that $P = MDM^{-1}$

(c) Compute $p_n(0, 0)$ and show that

$$|p_n(0, 0) - \pi(0)| \le \frac{4}{7(6)^n}.$$

In words, the convergence occurs exponentially fast.

2 Examples and Applications

2.1 Reversibility

A notion which is interesting in its own right and also helpful in computations is the following.

Definition 2.1. A probability distribution μ is said to be reversible with respect to a Markov chain with transition probabilities $p(i, j)$ if for all states i and j we have

$$\mu(i)p(i, j) = \mu(j)p(j, i).$$

It is much easier to find a reversible distribution (if it exists) than to find a stationary distribution. Reversible distributions are helpful because of the following.

Proposition 2.1. *A reversible probability distribution is stationary.*

Proof of Proposition 2.1. Assume μ is reversible with respect to the Markov chain with transition probabilities $p(i, j)$. By reversibility,

$$\sum_i \mu(i)p(i, j) = \sum_i \mu(j)p(j, i).$$

Note now that

$$\sum_i \mu(j)p(j, i) = \mu(j) \sum_i p(j, i) = \mu(j).$$

This shows that μ is a stationary distribution and completes the proof of Proposition 2.1.

As the next example shows the converse of Proposition 2.1 does not hold.

Example 2.1. Consider the following chain in $\{0, 1, 2\}$.

$$\begin{pmatrix} 0 & 1/4 & 3/4 \\ 3/4 & 0 & 1/4 \\ 1/4 & 3/4 & 0 \end{pmatrix}$$

It is easy to check that the unique stationary distribution π is given by $\pi(0) = \pi(1) = \pi(2) = \frac{1}{3}$. However,

$$\pi(0)p(0, 1) \neq \pi(1)p(1, 0).$$

Hence, π is not reversible.

2.2   Birth and Death Chains

Recall that a birth and death chain on the positive integers with a reflective boundary at 0 is given by

$$p(i, i + 1) = p_i \qquad p(i, i - 1) = q_i \qquad p(i, i) = r_i$$

where we assume that $p_i + q_i + r_i = 1$, $q_0 = 0$, $q_i > 0$ for $i \geq 1$ and $p_i > 0$ for $i \geq 0$. As noted before this chain is irreducible.

Note that if $|i - j| > 1$ then

$$\mu(i)p(i, j) = \mu(j)p(j, i)$$

is always true since both sides are 0. We now turn to the case $|i - j| = 1$. If μ is a reversible measure we must have for $i \geq 1$

$$\mu(i - 1)p(i - 1, i) = \mu(i)p(i, i - 1).$$

That is,

$$\mu(i - 1)p_{i-1} = \mu(i)q_i$$

so that

$$\mu(i) = \mu(i - 1)\frac{p_i}{q_{i+1}}.$$

Iterating this equality gives for $i \geq 1$

$$\mu(i) = \frac{p_{i-1}p_{i-2}\cdots p_0}{q_i q_{i-1}\cdots q_1}\mu(0).$$

We also want μ to be a probability distribution. Hence,

$$\sum_{i\geq 0}\mu(i) = 1.$$

That is,

$$\mu(0) + \mu(0)\sum_{i\geq 1}\frac{p_{i-1}p_{i-2}\cdots p_0}{q_i q_{i-1}\cdots q_1} = 1.$$

This equation has a strictly positive solution $\mu(0)$ if and only if

$$\sum_{i\geq 1}\frac{p_{i-1}p_{i-2}\cdots p_0}{q_i q_{i-1}\cdots q_1}$$

converges. We have proved the following.

Proposition 2.2. *A birth and death chain has a reversible distribution if and only if*

$$C = \sum_{i\geq 1}\frac{p_{i-1}p_{i-2}\cdots p_0}{q_i q_{i-1}\cdots q_1} < \infty. \tag{2.1}$$

If (2.1) holds, then the reversible distribution π is given by

$$\pi(0) = \frac{1}{C + 1}$$

and

$$\pi(i) = \frac{1}{C+1} \frac{p_{i-1} p_{i-2} \cdots p_0}{q_i q_{i-1} \cdots q_1} \text{ for } i \geq 1.$$

Condition (2.1) is sufficient to have a stationary distribution (why?). Is it necessary? It turns out that for an irreducible recurrent chain if we have μ such that for all j

$$\mu(j) = \sum_i \mu(i) p(i, j) \tag{2.2}$$

and

$$\sum_i \mu(i) = +\infty \tag{2.3}$$

then the chain cannot have a stationary distribution. This is an advanced result, see, for instance, Durrett (2010). If (2.1) fails, we have μ satisfying (2.2) and (2.3). Hence, there can be no stationary distribution.

Proposition 2.3. *The condition (2.1) is necessary and sufficient in order to have a stationary distribution. In particular, this shows that for birth and death chains the existence of a stationary distribution is equivalent to the existence of a reversible distribution.*

Recall that a birth and death chain is recurrent if and only if

$$\sum_{k \geq 1} \frac{q_k q_{k-1} \cdots q_1}{p_k p_{k-1} \cdots p_1} = +\infty.$$

On the other hand, condition (2.1) is necessary and sufficient to have positive recurrence (why?). Therefore, we have the following.

Proposition 2.4. *A birth and death chain is null recurrent if and only if*

$$\sum_{k \geq 1} \frac{q_k q_{k-1} \cdots q_1}{p_k p_{k-1} \cdots p_1} = +\infty \text{ and } \sum_{i \geq 1} \frac{p_{i-1} p_{i-2} \cdots p_0}{q_i q_{i-1} \cdots q_1} = +\infty.$$

2.3   The Simple Random Walk on the Half Line

Consider a simple random walk on the positive integers with a reflective boundary at 0. This is a particular birth and death chain. Given some p in $(0, 1)$ we have

$$p(i, i+1) = p \qquad p(i, i-1) = q = 1 - p \qquad p(i, i) = 0$$

for $i \geq 1$. We also assume that $q_0 = 0$, $p_0 = p$ and $r_0 = 1 - p$.

We will show that depending on p the random walk may be transient, null recurrent, or positive recurrent. Consider the Condition (2.1) in this particular case.

$$\sum_{i\geq 1} \frac{p_{i-1}p_{i-2}\cdots p_0}{q_i q_{i-1}\cdots q_1} = \sum_{i\geq 1}\frac{p^{i-1}}{q^i}.$$

Note that this series is convergent if and only if $p < q$ (i.e. $p < \frac{1}{2}$).

- Assume that $p < \frac{1}{2}$ then (2.1) holds and

$$C = \frac{p}{q-p}.$$

Hence, by Proposition 2.2 we have a reversible (and hence stationary) distribution given by

$$\pi(0) = \frac{1}{C+1} = 1 - \frac{p}{q}$$

and for $i \geq 1$

$$\pi(i) = (1 - \frac{p}{q})(\frac{p}{q})^i.$$

We also note that the random walk is positive recurrent (by Theorem 1.3) if and only if $p < \frac{1}{2}$.

- Assume that $p = \frac{1}{2}$. Since $p = q$ we get

$$\sum_{k\geq 1} \frac{q_k q_{k-1}\cdots q_1}{p_k p_{k-1}\cdots p_1} = \sum_{k\geq 1} 1 = +\infty.$$

Moreover,

$$\sum_{i\geq 1} \frac{p_{i-1}p_{i-2}\cdots p_0}{q_i q_{i-1}\cdots q_1} = \sum_{i\geq 1} 1 = +\infty.$$

Hence, by Proposition 2.4 the random walk is null recurrent for $p = \frac{1}{2}$.

- Assume that $p > \frac{1}{2}$ then

$$\sum_{k\geq 1} \frac{q_k q_{k-1}\cdots q_1}{p_k p_{k-1}\cdots p_1} = \sum_{k\geq 1}(\frac{q}{p})^k < +\infty$$

since $q < p$. Hence, the random walk is transient when $p > \frac{1}{2}$.

2.4   The Ehrenfest Chain

The Ehrenfest chain was introduced to model the process of heat exchange between two bodies that are in contact and insulated from the outside. The temperatures of the bodies are represented by the number of balls in two boxes. There are r balls labeled from 1 to r. Initially some of the balls are in box 1 and some of the balls are in box 2. At each step an integer between 1 and r is chosen at random and the ball with the corresponding label is moved from its box to the other box. Let X_n be the number of balls in box 1 at time n. The set of states is $\{0, 1, \ldots, r\}$. The transition probabilities are easy to compute:

$$p(i, i+1) = p_i = \frac{r-i}{r} \text{ and } p(i, i-1) = q_i = \frac{i}{r} \text{ for } 0 \le i \le r.$$

Hence, this is a birth and death chain with reflecting boundaries at 0 and r.

2.4.1   The Reversible Distribution

We use Proposition 2.2 to get

$$C = \sum_{i=1}^{r} \frac{p_{i-1} p_{i-2} \ldots p_0}{q_i q_{i-1} \ldots q_1} = \sum_{i=1}^{r} \frac{(r-i+1)(r-i+2)\ldots r}{i(i-1)\ldots 1} = \sum_{i=1}^{r} \binom{r}{i}.$$

Hence,

$$C + 1 = \sum_{i=0}^{r} \binom{r}{i}.$$

By Newton's formula

$$C + 1 = \sum_{i=0}^{r} \binom{r}{i} = (1+1)^r = 2^r.$$

Therefore,

$$\pi(0) = 2^{-r}$$

and for $i \ge 1$

$$\pi(i) = \binom{r}{i} 2^{-r}.$$

Hence, the reversible distribution is a binomial distribution with parameters r and $1/2$. In particular, the mean number of balls in box 1 at equilibrium is $r/2$. That is, at equilibrium we expect the temperatures of the two bodies to be the same.

2.4.2 Newton's Law of Cooling

We now derive Newton's law of cooling from the Ehrenfest chain. Let X_n be the number of balls in urn 1 at time n. We start by conditioning on X_n to get

$$E(X_{n+1}) = \sum_{k=0}^{r} E(X_{n+1}|X_n = k)P(X_n = k).$$

But

$$E(X_{n+1}|X_n = k) = (k+1)\frac{r-k}{r} + (k-1)\frac{k}{r} = k + \frac{r-2k}{r}.$$

Thus,

$$E(X_{n+1}) = \sum_{k=0}^{r}(k+\frac{r-2k}{r})P(X_n = k) = E(X_n)+1-\frac{2}{r}E(X_n) = (1-\frac{2}{r})E(X_n)+1.$$

Let Y_n be the difference of balls between the two urns at time n. That is, $Y_n = X_n - (r - X_n) = 2X_n - r$. We are going to compute the expected value of Y_n. We have

$$2E(X_{n+1}) - r = 2(1 - \frac{2}{r})E(X_n) + 2 - r = (1 - \frac{2}{r})(2E(X_n) - r).$$

Using that $Y_n = 2X_n - r$ in the preceding equality, we get

$$E(Y_{n+1}) = (1 - \frac{2}{r})E(Y_n).$$

We iterate the preceding equality to get

$$E(Y_n) = (1 - \frac{2}{r})^n E(Y_0).$$

This is the Newton's well-known law of cooling. We see that the expected temperature difference between the two bodies (i.e., $E(Y_n)$) decreases exponentially fast as a function of time n.

2.4.3 Reversibility Versus Irreversibility

The Ehrenfest model is supposed to model the transfer of heat from a warmer body into a colder body. This transfer, according to thermodynamics is irreversible: the heat transferred cannot go back. But in the Ehrenfest chain all possible transitions occur with probability one in a finite time since this is a positive recurrent chain (why?). So there seems to be a contradiction here. But we are going to show that the time it takes for the chain to make a transition opposite to the equilibrium is so large that if we look at this chain on a reasonable physical time scale it is extremely unlikely that we will see such a transition unless the chain is already very close to equilibrium.

Let $t(i, i + 1)$ be the random time it takes for the Ehrenfest chain to go from i to $i + 1$. We condition on the first transition. If the chain jumps to $i + 1$ then $t(i, i + 1) = 1$, if the chain jumps to $i - 1$ then the chain needs to go to i first and then to $i + 1$. Thus,

$$E(t(i, i + 1)) = p_i + q_i(E(t(i - 1, i) + E(t(i, i + 1))).$$

Solving for $E(t(i, i + 1))$ we get for $i \geq 1$

$$E(t(i, i + 1)) = 1 + \frac{q_i}{p_i} E(t(i - 1, i)).$$

The preceding formula holds for any birth and death chain with a reflecting boundary at the origin (i.e., $p_0 = 1$). Now we go back to the Ehrenfest chain. Recall that in this case

$$\frac{q_i}{p_i} = \frac{i}{r - i} \text{ for } 0 \leq i \leq r - 1.$$

We set $s_i = E(t(i, i + 1))$. The induction formula is then

$$s_i = 1 + \frac{i}{r - i} s_{i-1} \text{ for } 1 \leq i \leq r - 1.$$

Since there is a reflecting barrier at 0 we get $t(0, 1) = 1$ and $s_0 = 1$. Using the preceding formula with $r = 20$ gives $s_0 = 1, s_1 = 20/19, s_2 = 191/171, s_3 = 1160/969 \ldots$. The interesting part is that the first s_i are of the order of 1, s_{10} is about 4, s_{12} is about 10, s_{15} is about 150, and s_{19} is about 524,288. For $r = 20$ there is already a huge difference between the first s_i and the last ones. Note that 20 is a ridiculously small number in terms of atoms, we should actually be doing these computations for $r = 10^{23}$. For this type of r the difference between the first s_i and the last ones is inconceivably large. This shows that there is really no contradiction between the irreversibility in thermodynamics and the reversibility of the Ehrenfest model.

2.5 The First Appearance of a Pattern

Consider independent tosses of a coin that lands on heads with probability p and on tails with probability q.

Example 2.2. What is the expected number of tosses for T to appear? Denote by F_T the number of tosses for T to appear for the first time.

Since the tosses are independent, we get

$$P(F_T = n) = p^{n-1}q \text{ for } n \geq 1.$$

Hence,

$$E(F_T) = \sum_{n=1}^{\infty} np^{n-1}q.$$

Recall the geometric series for $|x| < 1$

$$\sum_{n=0}^{\infty} x^n = \frac{1}{1-x}.$$

Taking derivatives yield for $|x| < 1$

$$\sum_{n=1}^{\infty} nx^{n-1} = \frac{1}{(1-x)^2}.$$

Therefore,

$$E(F_T) = \sum_{n=1}^{\infty} np^{n-1}q = q\frac{1}{(1-p)^2} = \frac{1}{q}.$$

Example 2.3. What is the expected number of tosses for the pattern TT to appear?

We need to look at two consecutive tosses. Thus, we lose the independence between the outcomes. The number of steps to get TT is no longer geometric. To solve the problem we introduce a Markov chain. Let X_n be the last two outcomes after the nth toss. For $n \geq 2$, this is a Markov chain with four states $\{TT, TH, HT, HH\}$. It is also aperiodic since

$$P(X_{n+1} = TT | X_n = TT) = q > 0.$$

We know that a finite irreducible aperiodic chain has a unique stationary distribution that we denote by π. The convergence Theorem 1.1 holds and we have

$$\lim_{n\to\infty} P(X_n = TT) = \pi(TT).$$

In order for $X_n = TT$ we need tails on the nth and on the $(n-1)th$ tosses. This happens with probability q^2. Thus, $P(X_n = TT)$ is constant and $\pi(TT) = q^2$, see the problems for a different argument. According to Theorem 1.1

$$E(F_{TT}|X_0 = TT) = \frac{1}{q^2},$$

where F_{TT} is the time of first appearance of TT. Note now that to get TT we first need to get T and then we need to go from T to TT. If we couple two chains, one starting from T and one starting from TT they will agree from the first step on (why?). So the number of steps to go from T to TT is the same as the number of steps to go from TT to TT. Therefore,

$$E(F_{TT}) = E(F_T) + E(F_{TT}|X_0 = TT) = \frac{1}{q} + \frac{1}{q^2}.$$

Problems

1. Give an example of a stationary measure which is not reversible.

2. Assume X_n is an irreducible Markov chain with a reversible distribution π. Show that if $p(i, j) > 0$ then $p(j, i) > 0$.

3. Consider the chain with transition probabilities

$$\begin{pmatrix} 1/2 & 1/2 & 0 \\ 0 & 1/2 & 1/2 \\ 1/2 & 0 & 1/2 \end{pmatrix}.$$

(a) Use the preceding problem to show that this chain has no reversible distribution.
(b) What about a stationary distribution?

4. Take p in (0,1). Consider the one dimensional random walk on the positive integers

$$p(i, i+1) = p \qquad p(i, i-1) = 1 - p \text{ for } i \geq 1, \qquad p(0, 1) = 1.$$

that this chain is identical to the random walk on the half-line except that there
$p(0, 1) = p$.

(a) For what values of p is the chain positive recurrent? Find the stationary
distribution in this case.
(b) For what values of p is the chain null recurrent?

5. Consider the Ehrenfest chain with $r = 10^{23}$. Compute the expected time the
chain takes to return to the 0 state. (Use the stationary distribution.)

6. Decide whether the following birth and death chain is transient, positive
recurrent, or null recurrent. Let $p_0 = 1$ and

$$p_i = \frac{i}{4i+1} \qquad q_i = \frac{3i}{4i+1} \text{ for } i \geq 1.$$

7. We know that the random walk on the half line is null recurrent when $p = 1/2$.
In this problem we will show that a birth and death chain with $\lim_{i \to \infty} p_i = 1/2$
may be transient, null recurrent, or positive recurrent. Decide whether the following
birth and death chains are transient, positive recurrent, or null recurrent.

(a)

$$p_i = \frac{i+2}{2i+2} \qquad q_i = \frac{i}{2i+2}.$$

(b)

$$p_i = \frac{i+1}{2i+1} \qquad q_i = \frac{i}{2i+1}.$$

(c) Let $p_0 = 1$, $p_1 = q_1 = 1/2$ and

$$p_i = \frac{i-1}{2i} \qquad q_i = \frac{i+1}{2i} \text{ for } i \geq 2.$$

8. Consider an irreducible Markov chain with the property that $p(i, j) = p(j, i)$
for all i, j. Assume that there are only finitely many states. Find the reversible
distribution for this chain.

9. Let p be in $(0, 1)$. Consider N points on a circle. A random walk jumps to the
nearest point clockwise with probability p and counterclockwise with probability
$1 - p$.

(a) Show that this chain has a unique stationary distribution for all p in $(0, 1)$. Find
the stationary distribution.
(b) Show that this chain has a reversible distribution if and only if $p = \frac{1}{2}$.

10. Consider a birth and death chain with a reflecting barrier at 0 (i.e., $p_0 = 1$). Let $t(i, i + 1)$ be the random time it takes for chain to go from i to $i + 1$. Let $s_i = E(t(i, i + 1))$. We have shown that $s_0 = 1$ and for $i \geq 1$

$$s_i = 1 + \frac{q_i}{p_i} s_{i-1}.$$

(a) Consider a random walk on the half-line with a reflecting barrier at 0 and such that $p = 1/4$ and $q = 3/4$. Use the formula above to compute the expected time the walk takes to go from 9 to 10.
(b) How long does the walk take to go from 0 to 10?

11. Using the notation of the preceding problem show that for all $i \geq 1$ we have

$$s_i = 1 + \frac{q_i}{p_i} + \frac{q_i q_{i-1}}{p_i p_{i-1}} + \ldots + \frac{q_i q_{i-1} \cdots q_1}{p_i p_{i-1} \cdots p_1}.$$

12. Consider independent tosses of a coin that lands on heads with probability p and on tails with probability q. Let X_n be the last two outcomes after the nth toss. For $n \geq 2$, this is a Markov chain with four states $\{TT, TH, HT, HH\}$.

(a) Write the transition matrix P for the chain X_n.
(b) Let

$$\pi = (q^2, pq, pq, p^2).$$

Check that $\pi P = \pi$.

13. We use the notation of Example 2.3.

(a) Show that the expected number of tosses for the pattern HT to appear is $\frac{1}{pq}$.
(b) Compare (a) to Example 2.3.
(c) Set $p = q = \frac{1}{2}$. Do computer simulations to check the results in Example 2.3 and in (a).

Notes

We chose to omit a number of proofs in this chapter. The proofs of convergence and existence of stationary distributions are really analysis proofs and do not use many probability ideas. The reader may find the missing proofs in the references below. The Ehrenfest chain is analyzed in more detail in Bhattacharya and Waymire (1990). More in-depth exposition of the material can be found in the latter reference and in Karlin and Taylor (1975). Levin et al. (2008) provide a nice account of more modern topics such as mixing times.

References

Bhattacharya, R., Waymire, E.: Stochastic Processes with Applications. Wiley, New York (1990)

Durrett, R.: Probability: Theory and Examples, 4th edn. Cambridge University Press, Cambridge (2010)

Hoel, P., Port, S., Stone, C.: Introduction to Stochastic Processes. Houghton Mifflin, Boston (1972)

Karlin, S., Taylor, H.: A First Course in Stochastic Processes, 2nd edn. Academic, New York (1975)

Levin, D.A., Peres, Y., Wilmer, E.L.: Markov Chains and Mixing Times. American Mathematical Society, Providence (2008)

References

Chapter 7
The Poisson Process

In this chapter we introduce a continuous time stochastic process called the Poisson process. It is a good model in a number of situations and it has many interesting mathematical properties. There is a strong link between the exponential distribution and the Poisson process. This is why we start by reviewing the exponential distribution.

1 The Exponential Distribution

A random variable T is said to have an exponential distribution if it has a density $f(t) = \alpha e^{-\alpha t}$ for $t \geq 0$, where $\alpha > 0$ is the rate of the exponential distribution. In particular,

$$P(T > t) = \int_{t}^{\infty} f(s)ds = e^{-\alpha t} \text{ for } t \geq 0.$$

We may easily compute the mean and variance of T by integration by parts.

$$E(T) = \int_{0}^{\infty} tf(t)dt = \frac{1}{\alpha},$$

$$Var(T) = E(T^2) - E(T)^2 = \frac{1}{\alpha^2}.$$

The main reason the exponential distribution comes into play with Markov chains is the following property:

© Springer Science+Business Media New York 2014
R.B. Schinazi, *Classical and Spatial Stochastic Processes: With Applications to Biology*, DOI 10.1007/978-1-4939-1869-0_7

Proposition 1.1. *The exponential distribution has the following memoryless property. Assume T is a random variable with an exponential distribution. Then*

$$P(T > t + s | T > s) = P(T > t).$$

In words, waiting t units of time given that we have already waited s units of time is the same as waiting t units of time. That is, the system has no memory of having waited already s.

Proof of Proposition 1.1. By definition of conditional probability we have

$$P(T > t + s | T > s) = \frac{P(T > t + s; T > s)}{P(T > s)}.$$

But the event $\{T > t + s\}$ is included in the event $\{T > s\}$. Thus,

$$P(T > t + s | T > s) = \frac{P(T > t + s)}{P(T > s)} = \frac{e^{-\alpha(t+s)}}{e^{-\alpha s}} = e^{-\alpha t} = P(T > t).$$

This completes the proof of Proposition 1.1.

The exponential distribution is the only continuous distribution with the memoryless property, see Problem 1.1 for a proof.

We now turn to properties involving several independent exponential distributions. In particular, we are interested in the minimum of several independent exponential random variables. For instance, assume that T_1 and T_2 are exponentially distributed. We can define the minimum of T_1 and T_2. It is a new random variable which is T_1 when $T_1 < T_2$ and T_2 when $T_1 > T_2$ (the probability that $T_1 = T_2$ is 0). It turns out that the minimum of two independent exponential random variables is also an exponential variable whose rate is the sum of the two rates. We state the general result below.

Proposition 1.2. *Let $T_1, T_2, \ldots, T_n$ be independent exponential random variables with rates $\alpha_1, \alpha_2, \ldots, \alpha_n$. The random variable $\min(T_1, T_2, \ldots, T_n)$ is also exponentially distributed with rate $\alpha_1 + \alpha_2 + \cdots + \alpha_n$.*

Proof of Proposition 1.2. Observe that

$$P(\min(T_1, T_2, \ldots, T_n) > t) = P(T_1 > t; T_2 > t; \ldots; T_n > t)$$

and by independence we get

$$P(\min(T_1, T_2, \ldots, T_n) > t) = P(T_1 > t) P(T_2 > t) \ldots P(T_n > t) =$$

$$e^{-\alpha_1 t} e^{-\alpha_2 t} \ldots e^{-\alpha_n t} = e^{-(\alpha_1 + \alpha_2 + \ldots \alpha_n) t}.$$

This is enough to prove that $\min(T_1, T_2, \ldots, T_n)$ is exponentially distributed with rate $\alpha_1 + \alpha_2 + \cdots + \alpha_n$. This completes the proof of Proposition 1.2.

The following is another useful property.

Proposition 1.3. *Let* $T_1, T_2, \ldots, T_n$ *be independent exponential random variables with rates* $\alpha_1, \alpha_2, \ldots, \alpha_n$. *The probability that the minimum of the* T_i, $1 \le i \le n$, *is* T_k *for a given k is*

$$P(\min(T_1, T_2, \ldots, T_n) = T_k) = \frac{\alpha_k}{\alpha_1 + \alpha_2 + \cdots + \alpha_n}.$$

Proof of Proposition 1.3. Let $S_k = \min_{j \ne k} T_j$. Recall that the probability that two continuous and independent random variables be equal is zero. Thus,

$$P(\min(T_1, T_2, \ldots, T_n) = T_k) = P(T_k < S_k).$$

But according to the computation above, S_k is exponentially distributed with rate $\beta_k = \sum_{j \ne k} \alpha_j$ and S_k and T_k are independent. Thus,

$$P(T_k < S_k) = \int \int_{0 < t < s} \alpha_k e^{-\alpha_k t} \beta_k e^{-\beta_k s} \, dt \, ds = \frac{\alpha_k}{\alpha_k + \beta_k} = \frac{\alpha_k}{\alpha_1 + \alpha_2 + \cdots + \alpha_n}.$$

This completes the proof of Proposition 1.3.

We will use many times the following particular case of Proposition 1.3.

Corollary 1.1. *Let X and Y be two independent exponential random variables with rates a and b, respectively. We have that*

$$P(X < Y) = \frac{a}{a + b}.$$

The proof of Corollary 1.1 is an easy consequence of Proposition 1.3 and is left as an exercise.

Example 1.1. Assume that a system has two components A and B. The time it takes for components A and B to fail are exponentially distributed with rates 1 and 2, respectively. The components fail independently, and if one component fails the whole system fails. What is the expected time for the system to fail?

The time for the system to fail is the minimum of the failure times of components A and B. According to Proposition 1.2 this minimum is exponentially distributed with rate 3. Thus, the expected time for the system to fail is 1/3.

Problems

1. In this problem we show that the only continuous distribution with the memoryless property is the exponential. Let X be a random variable such that

$$P(X > t + s | X > t) = P(X > s) \text{ for all } t > 0, s > 0.$$

Let $u(t) = P(X > t)$.

(a) Show that for all $t > 0$ and $s > 0$ we have

$$u(t + s) = u(t)u(s).$$

(b) Show that if there is a such that $u(a) = 0$ then $u(t) = 0$ for all $t > 0$.

(c) Question (b) shows that we may consider u to be strictly positive. Define $e^{-\alpha} = u(1)$. Show that for every rational $r > 0$ we have

$$u(r) = e^{-\alpha r}.$$

(d) Use the continuity of the function u to show that for every real $t > 0$ we have

$$u(t) = e^{-\alpha t}.$$

2. Assume that a system has two components A and B. The time it takes for components A and B to fail are exponentially distributed with rates 2 and 3, respectively. The components fail independently and in order for the system to fail both components must fail.

(a) What is the distribution of the failure time for the system?
(b) What is the expected time for the system to fail?
(c) What is the probability that component A fails first?

3. Let S_n be a geometric random variable with success probability p_n. That is,

$$P(T_n = k) = (1 - p_n)^{k-1} p_n \text{ for } k = 1, 2, \ldots.$$

Assume that $\lim_{n \to \infty} n p_n = \alpha$. Compute

$$\lim_{n \to \infty} P(T_n / n > k).$$

4. Assume that the lifetime of a radio is exponentially distributed with mean 5 years.

(a) What is the probability that a new radio lasts more than 5 years?
(b) If the radio is already 5 years old, what is the probability that it lasts another 5 years?

5. Prove Corollary 1.1.

2 The Poisson Process

A stochastic process $(N(t))_{t \geq 0}$ is a collection of random variables $N(t)$. One of the most important stochastic processes is the so-called Poisson process. In many situations where we want to count the number of a certain type of random events happening up to time t the Poisson process turns out to be a good model. For instance, $N(t)$ may count the number of customers that visited a bank up to time t, or $N(t)$ could count the number of phone calls received at a home up to time t. In order to be Poisson a counting process needs to have the following properties.

Definition 2.1. A counting process $(N(t))_{t \geq 0}$ is said to be Poisson if it has the following properties:

(i) $N(0) = 0$
(ii) If $0 \leq s < t$, then $N(t) - N(s)$ has a Poisson distribution with parameter $\lambda(t - s)$. That is, for all integers $k \geq 0$

$$P(N(t) - N(s) = k) = \frac{e^{-\lambda(t-s)} \lambda^k (t - s)^k}{k!}.$$

(iii) $N(t)_{t \geq 0}$ has independent increments. That is, if $0 < t_1 < t_2 < \cdots < t_n$, then the random variables $N(t_2) - N(t_1), N(t_3) - N(t_2), \ldots, N(t_n) - N(t_{n-1})$ are independent.

Note that (ii) implies that the distribution of $N(t) - N(s)$ only depends on t and s through the difference $t - s$. The process $N(t)$ is said to have stationary increments. Property (iii) tells us that knowing what happened between times 0 and 1 does not give any information about what will happen between times 1 and 3. Hence, the events that we are counting need to be quite random for (iii) to be a reasonable hypothesis.

Example 2.1. Assume that phone calls arrive at a hospital at rate 2/min according to a Poisson process. What is the probability that 8 calls arrive during the first 4 min but no call arrives during the first minute?
    The event we are interested in is $\{N(4) = 8; N(1) = 0\}$. By (iii) we have

$$P(N(4) = 8; N(1) = 0) = P(N(1) = 0; N(4) - N(1) = 8)$$

By (iii)

$$P(N(1) = 0; N(4) - N(1) = 8) = P(N(1) = 0) P(N(4) - N(1) = 8).$$

By (ii),

$$P(N(4) - N(1) = 8) = P(N(3) = 8) = e^{-3\lambda} \frac{(3\lambda)^8}{8!} = e^{-8} \frac{6^8}{8!}.$$

Since, $P(N(1) = 0) = e^{-\lambda}$ we get

$$P(N(4) = 8; N(1) = 0) = e^{-\lambda}e^{-3\lambda}\frac{(3\lambda)^8}{8!} = e^{-8}\frac{6^8}{8!}.$$

We now give another characterization of the Poisson process. We first need a new notation. A function $r(h)$ is said to be $o(h)$ if

$$\lim_{h\to 0}\frac{r(h)}{h} = 0.$$

For instance, $r(h) = h^2$ is a $o(h)$ (why?). This is a convenient notation to avoid naming new functions in a computation.

Theorem 2.1. *Assume that $(N(t))_{t\geq 0}$ is a process on the positive integers with the following properties. There is a constant $\lambda > 0$ such that*

(a) $N(0) = 0$.
(b) $(N(t))_{t\geq 0}$ has stationary and independent increments.
(c) $P(N(h) = 1) = \lambda h + o(h)$.
(d) $P(N(h) \geq 2) = o(h)$.

Then $(N(t))_{t\geq 0}$ is a Poisson process with rate λ.

Theorem 2.1 is interesting in that it shows that a process with properties (a) through (d) is necessarily a Poisson process. There is no other choice!

Proof of Theorem 2.1. Since we are assuming $N(0) = 0$ and that $(N(t))_{t\geq 0}$ has stationary and independent increments we only need to check that $N(t)$ has a Poisson distribution for all $t > 0$.

Let $p_n(t) = P(N(t) = n)$ for $n \geq 0$. We first find p_0. We have

$$p_0(t + h) = P(N(t + h) = 0) = P(N(t) = 0; N(t + h) = 0)$$

where we are using that if $N(t + h) = 0$ then $N(s) = 0$ for all $s < t + h$. By independence of the increments we have

$$P(N(t) = 0; N(t + h) = 0) = P(N(t) = 0; N(t + h) - N(t) = 0)$$
$$= P(N(t) = 0)P(N(t + h) - N(t) = 0).$$

Since $N(t + h) - N(t)$ has the same distribution as $N(h)$ we get

$$P(N(t) = 0; N(t + h) = 0) = P(N(t) = 0)P(N(h) = 0).$$

Hence,

$$p_0(t + h) = p_0(t)p_0(h).$$

Observe now that by (c) and (d)

$$p_0(h) = P(N(h) = 0) = 1 - P(N(h) = 1) - P(N(h) \geq 2) = 1 - \lambda h - o(h) - o(h).$$

Since a linear combination of $o(h)$ is an $o(h)$ (why?) we get

$$p_0(h) = 1 - \lambda h + o(h) \qquad (2.1)$$

Therefore,

$$p_0(t + h) = p_0(t)(1 - \lambda h + o(h))$$

and

$$\frac{p_0(t + h) - p_0(t)}{h} = p_0(t)(-\lambda + \frac{o(h)}{h}).$$

Since $\lim_{h \to 0} \frac{o(h)}{h} = 0$, by letting h go to 0 we get

$$\frac{d}{dt} p_0(t) = -\lambda p_0(t).$$

Integrating this differential equation yields

$$p_0(t) = Ce^{-\lambda t}$$

where C is a constant. Note that $p_0(0) = P(N(0) = 0) = 1$ by (a). Hence, $C = 1$ and

$$p_0(t) = e^{-\lambda t}.$$

We now compute p_n for $n \geq 1$. If $N(t + h) = n$, then $N(t) = n$, $N(t) = n - 1$ or $N(t) < n - 1$. We now compute the corresponding probabilities.
    Using (b) we have

$$P(N(t) = n; N(t + h) = n) = p_n(t)p_0(h)$$

and

$$P(N(t) = n - 1; N(t + h) = n) = p_{n-1}(t)p_1(h).$$

Observe that in order for $N(t) < n-1$ and $N(t+h) = n$ there must be at least two events occurring between times t and $t+h$. Hence,

$$P(N(t) < n-1; N(t+h) = n) \leq P(N(t+h)-N(t) \geq 2) = P(N(h) \geq 2) = o(h).$$

Using the preceding three equations we get

$$p_n(t+h) = p_n(t)p_0(h) + p_{n-1}(t)p_1(h) + o(h).$$

By (2.1) and (c) we have

$$p_n(t+h) = p_n(t)(1 - \lambda h + o(h)) + (\lambda h + o(h))p_{n-1}(t) + o(h).$$

Note that multiplying an $o(h)$ by a bounded function (such as $p_n(t)$) yields another $o(h)$. Hence,

$$p_n(t+h) = p_n(t)(1 - \lambda h) + \lambda h p_{n-1}(t) + o(h).$$

and

$$\frac{p_n(t+h) - p_n(t)}{h} = -\lambda p_n(t) + \lambda p_{n-1}(t) + \frac{o(h)}{h}.$$

Letting h go to 0 we get for all $n \geq 1$

$$\frac{d}{dt} p_n(t) = -\lambda p_n(t) + \lambda p_{n-1}(t).$$

We now transform this differential equation by multiplying both sides by $e^{\lambda t}$.

$$e^{\lambda t} \frac{d}{dt} p_n(t) = -\lambda e^{\lambda t} p_n(t) + \lambda e^{\lambda t} p_{n-1}(t).$$

So

$$e^{\lambda t} \frac{d}{dt} p_n(t) + \lambda e^{\lambda t} p_n(t) = \lambda e^{\lambda t} p_{n-1}(t).$$

Observe that the l.h.s. is exactly the derivative of $e^{\lambda t} p_n(t)$. Hence,

$$\frac{d}{dt}(e^{\lambda t} p_n(t)) = \lambda e^{\lambda t} p_{n-1}(t) \text{ for all } n \geq 1 \qquad (2.2)$$

We use (2.2) to compute $p_n(t)$. We will prove by induction that

$$p_n(t) = e^{-\lambda t} \frac{(\lambda t)^n}{n!} \text{ for all } n \geq 0 \tag{2.3}$$

We know that $p_0(t) = e^{-\lambda t}$. Hence, (2.3) holds for $n = 0$. Assume now that (2.3) holds for n and use (2.3) to get

$$\frac{d}{dt}(e^{\lambda t} p_{n+1}(t)) = \lambda e^{\lambda t} p_n(t) = \lambda e^{\lambda t} e^{-\lambda t} \frac{(\lambda t)^n}{n!}.$$

Hence,

$$\frac{d}{dt}(e^{\lambda t} p_{n+1}(t)) = \lambda \frac{(\lambda t)^n}{n!} = \frac{\lambda^{n+1}}{n!} t^n.$$

We integrate both sides with respect to t to get

$$e^{\lambda t} p_{n+1}(t) = \frac{\lambda^{n+1}}{n!} \frac{t^{n+1}}{n+1} + C = \frac{(\lambda t)^{n+1}}{(n+1)!} + C.$$

Note that $p_{n+1}(0) = P(N(0) = n + 1) = 0$. Therefore, $C = 0$ and we have

$$p_{n+1}(t) = e^{-\lambda t} \frac{(\lambda t)^{n+1}}{(n+1)!}.$$

Therefore, (2.3) holds for $n + 1$ and this formula is proved by induction. This completes the proof of Theorem 2.1.

The next result is important for at least two reasons. It shows that the Poisson process is closely related to the exponential distribution and it gives a method to construct a Poisson process.

(I) Let $\tau_1, \tau_2 \ldots$ a sequence of independent random variables with the same rate λ exponential distribution. Let $T_0 = 0$ and for $n \geq 1$ let

$$T_n = \tau_1 + \tau_2 + \cdots + \tau_n.$$

(II) For $t \geq 0$ let

$$N(t) = \max\{n \geq 0 : T_n \leq t\}.$$

In words, τ_1 is the time of the first event and for $n \geq 2$, τ_n is the time between the $(n - 1) - th$ event and the n-th event. For $n \geq 1$, T_n is the time of the n-th event. For $t \geq 0$, $N(t)$ counts the number of events occurring by time t.

Theorem 2.2. *A counting process* $(N(t))_{t\geq 0}$ *defined by (I) and (II) is necessarily a Poisson process with rate* λ.

Proof of Theorem 2.2. We first show that the distribution of $N(t)$ is Poisson. Note that if $N(t) = n$ then the n-th event has occurred by time t but the $(n + 1)$-th has not. Hence,

$$P(N(t) = n) = P(T_n \leq t < T_{n+1}).$$

Since T_n is the sum of n i.i.d. exponential random variables it has a Γ distribution and its density is

$$f(s) = \frac{\lambda^n s^{n-1}}{(n-1)!} e^{-\lambda s}.$$

Note also that

$$T_{n+1} = T_n + \tau_{n+1}$$

and that T_n and τ_{n+1} are independent. The joint density of (T_n, τ_{n+1}) is

$$g(s, u) = \frac{\lambda^n s^{n-1}}{(n-1)!} e^{-\lambda s} \lambda e^{-\lambda u}.$$

Therefore,

$$P(N(t) = n) = P(T_n \leq t < T_n + \tau_{n+1}) = \int\int_{s \leq t < s+u} \frac{\lambda^n s^{n-1}}{(n-1)!} e^{-\lambda s} \lambda e^{-\lambda u} ds\, du.$$

Integrating first in u yields

$$\int_{u>t-s} \lambda e^{-\lambda u} du = e^{-\lambda(t-s)}.$$

Hence,

$$P(N(t) = n) = \int_0^t \frac{\lambda^n s^{n-1}}{(n-1)!} e^{-\lambda s} e^{-\lambda(t-s)} ds = e^{-\lambda t} \int_0^t \frac{\lambda^n s^{n-1}}{(n-1)!} ds = e^{-\lambda t} \lambda^n \frac{t^n}{n!}.$$

This shows that $N(t)$ has a Poisson distribution with parameter $N(t)$. In order to prove that $(N(t))_{t\geq 0}$ is actually a Poisson process we still need to show that the increments of this process are stationary and independent. We will not quite prove that. The formal proof is a little involved, see, for instance, Durrett (2010). Instead, we will show that the location of the first event after time t is independent of $N(t)$. This is really the critical part of the proof and it is the following lemma. First, a new notation. Recall that up to time t there are $N(t)$ events. So the last event before time t occurs at time $T_{N(t)}$ and the first event after time t occurs at time $T_{N(t)+1}$.

Lemma 2.1. *Assuming hypotheses (I) and (II) the random variables $T_{N(t)+1} - t$ and $N(t)$ are independent. That is, the location of the first event after time t is independent of $N(t)$. Moreover, the random variable $T_{N(t)+1} - t$ is exponentially distributed with rate λ.*

The event $N(t) = n$ is the same as the event $T_n \leq t < T_{n+1}$. Hence,

$$P(T_{n+1} \geq t + v; N_t = n) = P(T_{n+1} \geq t + v; T_n \leq t).$$

Using again the joint density of (T_n, τ_{n+1}) we have

$$P(T_{n+1} \geq t + v; T_n \leq t) = \int \int_{s \leq t; s+u \geq t+v} \frac{\lambda^n s^{n-1}}{(n-1)!} e^{-\lambda s} \lambda e^{-\lambda u} ds\, du.$$

Integrating first in u yields

$$P(T_{n+1} \geq t + v; T_n \leq t) = \int_{s \leq t} \frac{\lambda^n s^{n-1}}{(n-1)!} e^{-\lambda s} e^{-\lambda(t+v-s)} ds$$

$$= e^{-\lambda(t+v)} \int_{s \leq t} \frac{\lambda^n s^{n-1}}{(n-1)!} ds = e^{-\lambda(t+v)} \frac{\lambda^n t^n}{n!}.$$

Using now that

$$P(N(t) = n) = e^{-\lambda t} \frac{\lambda^n t^n}{n!}$$

we have

$$P(T_{n+1} - t \geq v; N(t) = n) = P(T_{N(t)+1} - t \geq v; N_t = n) = e^{-\lambda v} P(N(t) = n).$$

That is, the probability of the intersection

$$\{T_{N(t)+1} - t \geq v\} \cap \{N_t = n\}$$

is the product of the corresponding probabilities. Hence, the random variables $T_{N(t)+1} - t$ and $N(t)$ are independent. The time it takes for the first event after time t to occur (i.e., $T_{N(t)+1} - t$) is independent of what occurred before time t. This is a consequence of the memory less property of the exponential distribution. This completes the proof of Lemma 2.1 and of Theorem 2.2.

We now show that the converse of Theorem 2.2 is also true.

Theorem 2.3. *Assume that $(N(t))_{t \geq 0}$ is a Poisson process. Then the inter arrival times between two events are independent and exponentially distributed.*

Proof of Theorem 2.3. We will give an informal argument that should convince the reader that the Theorem is true. A formal proof can be found in a more advanced text such as Bhattacharya and Waymire (1990).

We are given a Poisson process $(N(t))_{t \geq 0}$. Let τ_1 be the time of the first event, let τ_2 be the time elapsed between the first and second event and so on. Note that

$$P(\tau_1 > t) = P(N(t) = 0) = e^{-\lambda t}.$$

This shows that τ_1 is exponentially distributed with rate λ. We now show that τ_2 has the same distribution as τ_1 and that these two random variables are independent.

Conditioning on the first arrival time we have

$$P(\tau_2 > t; \tau_1 > s) = \int_s^\infty P(\tau_2 > t | \tau_1 = u) P(\tau_1 = u) du$$

$$= \int_s^\infty P(\tau_2 > t | \tau_1 = u) \lambda e^{-\lambda u} du.$$

Note that

$$P(\tau_2 > t | \tau_1 = u) = P(N(t + u) - N(u) = 0 | \tau_1 = u).$$

Since $(N(t))_{t \geq 0}$ has independent increments we can argue that the events $\{N(t + u) - N(u) = 0\}$ and $\{\tau_1 = u\}$ are independent. This is so because the first event depends on what happens between times u and $t + u$ while the second event depends on what happens between times 0 and u. Hence,

$$P(\tau_2 > t | \tau_1 = u) = P(N(t + u) - N(u) = 0 | \tau_1 = u)$$
$$= P(N(t + u) - N(u) = 0) = P(N(t) = 0) = e^{-\lambda t}.$$

Going back to the integral we get

$$P(\tau_2 > t; \tau_1 > s) = \int_s^\infty e^{-\lambda t} \lambda e^{-\lambda u} du = e^{-\lambda t} e^{-\lambda s}.$$

This shows that τ_1 and τ_2 are independent and that they are exponentially distributed. This argument can be easily generalized to any finite number of τ_i's. This completes the proof of Theorem 2.3.

Example 2.2. Consider phone calls arriving at a hospital at rate 2/min according to a Poisson process.

(a) What is the expected time of the third call?
       The third call occurs at time T_3. We have

$$T_3 = \tau_1 + \tau_2 + \tau_3.$$

We now use that each interarrival time has mean $1/\lambda = 1/2$. Thus,

$$E(\tau_3) = 3/2.$$

(b) What is the probability that the third call arrives after 2 min?

We want to compute the probability of event $\{T_3 > 2\}$. This is the same as the event $\{N(2) < 3\}$. Since $N(2)$ follows a Poisson distribution with parameter $2\lambda = 4$ we have

$$P(N(2) < 3) = P(N(2) = 0) + P(N(2) = 1) + P(N(2) = 2)$$

$$= e^{-4}(1 + 4 + \frac{4^2}{2!}) = 13e^{-4}.$$

Example 2.3. Consider the following problem. Assume that calls arrive to a fire station according to a Poisson process with rate 20/h. Assume also that only about 20 % of these calls are emergency calls. Consider the process of emergency calls only. Is this still a Poisson process? If this is the case at what rate do the emergency calls arrive? These questions are answered by the following theorem.

Theorem 2.4. *Let $(N(t)_{t\geq0}$ be a Poisson process with rate λ. Occurrences of the Poisson process may be of type 1 or type 2 with probability p and $1 - p$, respectively, independently of all other events. Let $N_1(t)$ and $N_2(t)$ be the processes of occurrences of type 1 and type 2, respectively. Then $N_1(t)$ and $N_2(t)$ are independent Poisson processes with rates λp and $\lambda(1 - p)$, respectively.*

It is not so surprising that $N_1(t)$ and $N_2(t)$ are Poisson processes with the stated rates. What might seem surprising at first glance is that they are independent.

Going back to Example 2.3. We have $\lambda = 20$, $p = 1/5$. Hence, the emergency calls form a Poisson process with rate $\lambda p = 4$/h.

Proof of Theorem 2.4. We start by computing the joint distribution of $(N_1(t), N_2(t))$. Note that

$$N(t) = N_1(t) + N_2(t).$$

Let k and n be positive integers.

$$P(N_1(t)=k; N_2(t) = n)=P(N_1(t)=k; N_2(t)=n|N(t)=k + n)P(N(t)=k + n).$$

Given $N(t) = k +n$, since occurrences are of type 1 or 2 independently of all other events, the number of type 1 events follows a binomial with parameters $n + k$ and p. Thus,

$$P(N_1(t) = k; N_2(t) = n|N(t) = k + n) = \binom{n + k}{k} p^k(1 - p)^n$$

and

$$P(N_1(t) = k; N_2(t) = n) = \binom{n+k}{k} p^k (1-p)^n \frac{(\lambda t)^{k+n}}{(k+n)!} e^{-\lambda t}.$$

A little algebra gives

$$P(N_1(t) = k; N_2(t) = n) = \frac{(\lambda p t)^k}{k!} e^{-\lambda p t} \frac{(\lambda(1-p)t)^n}{n!} e^{-\lambda(1-p)t}. \tag{2.4}$$

We may now compute the distribution of $N_1(t)$:

$$P(N_1(t) = k) = \sum_{n \geq 0} P(N_1(t) = k; N_2(t) = n)$$

$$= \sum_{n \geq 0} \frac{(\lambda p t)^k}{k!} e^{-\lambda p t} \frac{(\lambda(1-p)t)^n}{n!} e^{-\lambda(1-p)t}.$$

We sum the series to get that

$$P(N_1(t) = k) = \frac{(\lambda p t)^k}{k!} e^{-\lambda p t}.$$

This shows that $N_1(t)$ has a Poisson distribution with rate $\lambda p t$. A similar computation shows that

$$P(N_2(t) = n) = \frac{(\lambda(1-p)t)^n}{n!} e^{-\lambda(1-p)t}.$$

Hence, using (2.4)

$$P(N_1(t) = k; N_2(t) = n) = P(N_1(t) = k) P(N_2(t) = n)$$

showing that the two processes are independent.

We now prove that $(N_1(t))_{t \geq 0}$ has stationary increments. We have for $s < t$ and any positive integer k

$$P(N_1(t) - N_1(s) = k) = \sum_{n \geq 0} P(N_1(t) - N_1(s) = k; N(t) - N(s) = k + n).$$

We use again that given $N(t) - N(s) = k + n$, the distribution of $N_1(t) - N_1(s)$ is a binomial distribution with parameters $k + n$ and p. Thus,

$$P(N_1(t) - N_1(s) = k) = \sum_{n \geq 0} \binom{n+k}{k} p^k (1-p)^n P(N(t) - N(s) = k + n).$$

Since $N(t) - N(s)$ has a Poisson distribution with rate $\lambda(t - s)$ we get

$$P(N_1(t) - N_1(s) = k) = \sum_{n \geq 0} \binom{n + k}{k} p^k (1 - p)^n e^{-\lambda(t-s)} \frac{(\lambda(t - s))^{k+n}}{(k + n)!},$$

so

$$P(N_1(t) - N_1(s) = k) = p^k e^{-\lambda(t-s)} \frac{(\lambda(t - s))^k}{k!} \sum_{n \geq 0} (1 - p)^n \frac{(\lambda(t - s))^n}{n!}.$$

After summing the series we have

$$P(N_1(t) - N_1(s) = k) = e^{-\lambda p(t-s)} \frac{(\lambda p(t - s))^k}{k!}.$$

That is, $N_1(t) - N_1(s)$ has the same distribution as $N_1(t - s)$. Hence, $(N_1(t))_{t \geq 0}$ has stationary increments.

Note that the distribution of $N_1(t) - N_1(s)$ depends only on the distribution of $N(u)$ for u in $[s, t]$. So, using the fact that the increments of $N(t)$ are independent, we see that the increments of $N_1(t)$ are also independent.

We have checked (i), (ii), and (iii) from Definition 2.1 for $N_1(t)$. Thus, this is a Poisson process. The process $N_2(t)$ has the same properties as the process $N_1(t)$. Hence, $N_1(t)$ and $N_2(t)$ are independent Poisson processes and the proof of Theorem 2.4 is complete.

2.1   Application: Influenza Pandemics

Since 1700 there have been about ten worldwide severe influenza pandemics. The dates are 1729, 1781, 1799, 1830, 1847, 1889, 1918, 1957, 1968, 1977. There are not really enough data points to decide whether the times between pandemics are exponentially distributed. Even so a Poisson process gives a model that does not appear unreasonable and for which we may compute probabilities. The average interarrival time between two pandemics is about 25 years. We take 1700 as time 0 and we use a Poisson process to model the number of pandemics with $\lambda = 1/25$.

The main question of interest in this section is to compute the probability of long periods without pandemics. One of the reasons for the anxiety about the 2008 influenza season was that the last pandemic had occurred 30 or 40 years earlier and therefore we were due for a new pandemic. Or were we?

With the notation of this section, there are $N(t)$ pandemics up to time t and the first pandemic after time t occurs at time $T_{N(t)+1}$. The time between t and the next pandemic (i.e., $T_{N(t)+1} - t$) was shown to be exponentially distributed with rate λ in Lemma 2.1.

The probability that the first pandemic after time t be at least twice the mean time $1/\lambda$ is therefore

$$\int_{2/\lambda}^{+\infty} \lambda \exp(-\lambda x)dx = \exp(-2) \sim 14\%.$$

Similarly the probability that the first pandemic after time t be at least three times the mean is $\exp(-3) \sim 5\%$. Hence, under this model it is not unlikely that starting today we have no pandemic for another 50 years or even 75 years.

Problems

1. Show that if $N(t)$ is a Poisson process and $s < t$ then

$$P(N(s) = k | N(t) = n) = \binom{n}{k} (\frac{s}{t})^k (1 - \frac{s}{t})^{n-k}$$

for $k = 0, \ldots, n$.

2. Let $N(t)$ be a Poisson process with rate λ. Compute $E(N(t)N(t+s))$.

3. Assume $N_1(t)$ and $N_2(t)$ are independent Poisson processes with rates λ_1 and λ_2. Show that $N_1(t) + N_2(t)$ is a Poisson process with rate $\lambda_1 + \lambda_2$.

4. Assume that $N(t)$ is a Poisson process and let τ_1 be the time of the first event. Prove that if $s \leq t$ then

$$P(\tau_1 < s | N(t) = 1) = \frac{s}{t}.$$

5. Assume $N(t)$ is a Poisson process with rate λ, and $(Y_i)_{i \geq 1}$ are i.i.d. random variables. Assume also that $N(t)$ and the Y_i are independent. Define the *compound Poisson process* by

$$X(t) = \sum_{i=1}^{N(t)} Y_i.$$

Show that

$$E(X(t)) = \lambda t E(Y_1).$$

6. Assume that certain events occur according to a Poisson process with rate 2/h.

(a) What is the probability that no event occurs between 8:00 and 9:00 PM?

(b) What is the expected time at which the fourth event occurs?

(c) What is the probability that two or more events occur between 8:00 and 9:00 PM?

7. Emails arrive according to a Poisson process with rate 20/day. The proportion of junk email is 90 %.

(a) What is the probability of getting at least one non-junk email today?

(b) What is the expected number of junk emails during 1 week?

8. Customers arrive at a bank according to a Poisson process with rate 10/h.

(a) Given that exactly two customers came the first hour, what is the probability they both arrived during the first 20 min?

(b) What is the probability that the fourth customer arrives after 1 h?

9. The following ten numbers are simulations of exponential random variables with rate 1: 4.78, 1.05, 0.92, 2.21, 3.22, 2.21, 4.6, 5.28, 1.97, 1.39. Use these numbers to simulate a rate 1 Poisson process.

10. Consider a Poisson process $(N(t))_{t \geq 0}$ with rate λ. Let T_n be the arrival time of the n-th event for some $n \geq 1$.

(a) Show that

$$P(N(t) < n) = P(T_n > t).$$

(b) Show that

$$P(T_n > t) = \sum_{k=0}^{n-1} e^{-\lambda t} \frac{(\lambda t)^k}{k!}.$$

(c) Show that the density of the distribution of T_n is

$$f(t) = \frac{\lambda^n t^{n-1}}{(n-1)!} e^{-\lambda t}.$$

(Use (b) and recall that $\frac{d}{dt} P(T_n \leq t) = f(t)$.)

11. Recall that the distribution function of a random variable X is defined by

$$F_X(x) = P(X \leq x).$$

(a) Let T be exponentially distributed with rate λ. Show that for $t \geq 0$ we have

$$F_T(t) = 1 - e^{-\lambda t}.$$

(b) Find the inverse function F_T^{-1}.

(c) Let U be a uniform random variable on $[0, 1]$. Let $Y = F_T^{-1}(U)$. Show that

$$F_Y(y) = P(Y \le y) = F_T(y).$$

That is, the random variable

$$Y = -\frac{1}{\lambda} \ln(1 - U)$$

is exponentially distributed with rate λ.

12. (a) Simulate 10 independent exponential observations with rate 1/25. (See the method in problem 11.)

(b) Use these observations to simulate 10 pandemic dates. Take the origin to be year 1700.

(c) Compare your simulation to the real data.

13. Let X and Y be independent random variables with rates a and b, respectively.

(a) Show that

$$P(0 < X < h) = ah + o(h).$$

(b) Show that

$$P(\{0 < X < h\} \cap \{0 < Y < h\}) = o(h).$$

(c) Show that

$$P(\{X > h\} \cap \{Y > h\}) = 1 - (a + b)h + o(h).$$

14. (a) Show that a linear combination of $o(h)$ is $o(h)$.

(b) Show that a product of $o(h)$ and a bounded function is $o(h)$.

Notes

We give a very short introduction of the Poisson process. Kingman (1993) is an excellent reference on the subject. At a higher mathematical level Durrett (2010) and Bhattacharya and Waymire (1990) are good references for the Poisson process and more generally for probability and stochastic processes.

References

Bhattacharya, R.N., Waymire, E.C.: Stochastic Processes with Applications. Wiley, New York (1990)

Durrett, R.: Probability: Theory and Examples, 4th edn. Cambridge University Press, Cambridge (2010)

Kingman, J.F.C.: Poisson Processes. Oxford University Press, New York (1993)

Chapter 8
Continuous Time Branching Processes

We introduce continuous time branching processes. The main difference between discrete and continuous branching processes is that births and deaths occur at random times for continuous time processes. Continuous time branching processes have the Markov property if (and only if) birth and death times are exponentially distributed. We will use several properties of the exponential distribution.

1 A Continuous Time Binary Branching Process

We define a continuous time binary branching process by the following rules. Each individual gives birth to a new individual at rate λ or dies at rate 1. Individuals are independent of each other.

More precisely, each individual in the population has two independent exponential random variables attached to it. One random variable has rate λ, the other one has rate 1. If the rate λ exponential random variable happens before the rate 1 exponential, then the individual is replaced by two individuals. If the rate 1 exponential random variable happens before the rate λ exponential then the individual dies with no offspring. Every new individual gets two independent exponential random variables attached to it (one with rate λ and the other with rate 1) and so on.

The number of individuals at time t is denoted by Z_t. We start the process with $Z_0 = 1$.

The following is the main result of this section.

© Springer Science+Business Media New York 2014
R.B. Schinazi, *Classical and Spatial Stochastic Processes: With Applications to Biology*, DOI 10.1007/978-1-4939-1869-0_8

Theorem 1.1. *Consider a binary branching process* $(Z_t)_{t\geq0}$ *with birth rate* λ *and death rate 1. The process has a strictly positive probability of surviving if and only if* $\lambda > 1$. *Moreover, for any* $\lambda > 0$ *we have*

$$E(Z_t) = e^{(\lambda-1)t}.$$

Proof of Theorem 1.1. Consider the process Z_t at integer times n. Assume that at time $n - 1$ there are $Z_{n-1} = j \geq 0$ individuals. If $j = 0$, then $Z_t = 0$ for all $t > n - 1$. If $j \geq 1$, then label each of the individuals present at time $n - 1$. For $1 \leq k \leq j$ let Y_k be the number of descendants (which can possibly be 0) of the kth individual after one unit time. If the kth individual has not undergone a split between times $n - 1$ and n it means that this individual is still present at time n and we let $Y_k = 1$. Note that the $(Y_k)_{1\leq k\leq j}$ are independent and identically distributed. Moreover,

$$Z_n = \sum_{k\geq0}^{j} Y_k \text{ for all } n \geq 1.$$

Thus, Z_n is a discrete time branching process. Observe also that $Z_t = 0$ for some $t > 0$ if and only if $Z_n = 0$ for some $n \geq 1$ (why?). Hence,

$$P(Z_t \neq 0, \text{ for all } t > 0) = P(Z_n \neq 0, \text{ for all } n > 0).$$

In other words, the problem of survival for the continuous time branching process is equivalent to the problem of survival for the corresponding discrete time process. We know that the discrete time process survives if and only if $E(Y_1) > 1$. The only difficulty is that we do not know the distribution of Y_1. It is however possible to compute the expected value of Y_1 without computing the distribution of Y_1. We do this now.

Denote the expected number of particles at time t by

$$M(t) = E(Z_t|Z_0 = 1).$$

Next we derive a differential equation for $M(t)$. We condition on what happens between times 0 and h where h is small (we will let h go to 0). At time 0 we have a single individual. There are three possibilities.

(1) The individual gives birth between times 0 and h. This happens with probability $\lambda h + o(h)$ (why?).
(2) The individual dies. This happens with probability $h + o(h)$.
(3) Nothing happens at all with probability $1 - (\lambda + 1)h + o(h)$.

We have

$$M(t + h) = \lambda h(2M(t)) + (1 - (\lambda + 1)h)M(t).$$

This is so because of our three possibilities above. If there is a birth, then at time h there are two individuals and each one founds a new process that runs until time $t + h$. If the individual dies, then $M(t + h) = 0$ and we can ignore this term. If nothing happens by time h, then we have a single individual at time h.

Therefore,

$$\frac{M(t + h) - M(t)}{h} = \lambda M(t) - M(t) + \frac{o(h)}{h}.$$

Letting h go to 0 we have

$$M'(t) = (\lambda - 1)M(t).$$

Integrating this differential equation with the initial value $M(0) = 1$ gives

$$M(t) = e^{(\lambda-1)t}.$$

So $E(Z_1) = E(Y_1) = e^{\lambda-1}$. Note that $E(Y_1) > 1$ if and only if $\lambda > 1$. That is, the process Z_t survives forever with positive probability if and only if $\lambda > 1$. This completes the proof of Theorem 1.1.

Remark. Consider a continuous time branching process with $Z_0 = 1$. Say that $Z_1 = 3$. Each one of these three individuals has appeared at a different time between times 0 and 1. However, we claim that $(Z_n)_{n \geq 0}$ is a discrete time branching process. In particular, we claim that each one of these three individuals starts processes at time 1 that have the same distribution. This is so because of the memoryless property of the exponential distribution. It does not matter how old the individual is at time 1. All that matters is that the individual is present at time 1.

Problems

1. In this problem we give a more general definition of a continuous time branching process.

After an exponential time with parameter a a given individual is replaced by $k \geq 0$ individuals with probability f_k, where $(f_k)_{k \geq 0}$ is a probability distribution on the positive integers. Hence, an individual dies leaving no offspring with probability f_0. Each individual evolves independently of the others. Denote by Z_t the number of individuals at time t. Let $M(t) = E(Z_t)$.

(a) Show that

$$M(t + h) = ah \sum_{k=0}^{\infty} k f_k M(t) + (1 - ah)M(t) + o(h).$$

(b) Show that

$$\frac{d}{dt}M(t) = a(c-1)M(t)$$

where c is the mean offspring

$$c = \sum_{k=0}^{\infty} k f_k.$$

(c) Show that

$$M(t) = e^{a(c-1)t}.$$

(d) Show that the process survives if and only if $c > 1$.

2. Consider a particular case of the process in problem 1 with $f_k = \frac{1}{2}\frac{1}{2^k}$ for $k \geq 0$.

(a) Does this process survive?
(b) Compute $E(Z_t | Z_0 = 1)$.

3. In this problem we show that the binary branching process is a particular case of the branching process defined in problem 1. We assume that the birth rate is λ and the death rate 1. We need to find a and the probability distribution $(f_k)_{k \geq 0}$.

(a) Show that after an exponential time with rate $a = \lambda + 1$ an individual in the binary branching process is replaced by 0 or 2 individuals.
(b) Show that the probability for an individual to be replaced by 0 individuals is $f_0 = \frac{1}{\lambda+1}$ and to be replaced by 2 individuals is $f_2 = \frac{\lambda}{\lambda+1}$.

4. Consider a continuous time binary branching starting with a single individual. We assume that the birth rate is λ and the death rate 1.

(a) Show that the probability that there is a birth between times 0 and h is $1 - e^{-\lambda h}$.
(b) Show that

$$1 - e^{-\lambda h} = 1 - \lambda h + o(h).$$

(c) Show that the probability that there is no birth and no death between times 0 and h is $e^{-(\lambda+1)h}$.
(d) Show that

$$e^{-(\lambda+1)h} = 1 - (\lambda + 1)h + o(h).$$

2   A Model for Immune Response

We introduce a stochastic model designed to test the following hypothesis: can a pathogen escape the immune system only because of its high probability of mutation? For instance, the HIV virus is known to mutate at a very high rate and it is thought that it overwhelms the human immune system because of this high mutation rate.

We now describe the model. We have two parameters $\lambda > 0$ and r in $[0, 1]$.

Births. We start the model with a single pathogen at time zero. Each pathogen gives birth to a new pathogen at rate λ. That is, a random exponential time with rate λ is attached to each pathogen. These exponential times are independent of each other. When the random exponential time occurs a new pathogen is born. The new pathogen has the same type as its parent with probability $1 - r$. With probability r, a mutation occurs, and the new pathogen has a different type from all previously observed pathogens. For convenience, we say that the pathogen present at time zero has type 1, and the kth type to appear will be called type k. Note that we assume the birth rate λ to be the same for all types and we therefore ignore selection pressures.

Deaths. Each pathogen that is born is killed after an exponentially distributed time with mean 1. When a pathogen is killed, all pathogens of the same type are killed simultaneously. In other words, each pathogen is born with an exponential clock which, when it goes off, kills all pathogens of its type.

The rule that all pathogens of the same type are killed simultaneously is supposed to mimic the immune response. Note that types that have large numbers of pathogens are more likely to be targeted by the immune system and eliminated.

We start with the following result.

Proposition 2.1. *If $\lambda \leq 1$, then the pathogens die out for all r in $[0, 1]$.*

In words, even if the probability of mutation is very high the pathogens cannot survive for a birth rate λ less than 1. Proposition 2.1 is actually included in Theorem 2.1 below. We include Proposition 2.1 because its proof is elementary and uses an important technique, the so-called *coupling* technique.

Proof of Proposition 2.1. Let $(X_t)_{t \geq 0}$ be the number of pathogens at time $t \geq 0$. We start the process with one pathogen so that $X_0 = 1$. Consider now the process $(Z_t)_{t \geq 0}$. We let $Z_0 = 1$ and let $(Z_t)_{t \geq 0}$ evolve as $(X_t)_{t \geq 0}$ with the same birth and death rates. The only difference between the two processes is the following. Just one pathogen dies at a time for the process $(Z_t)_{t \geq 0}$ while for the process $(X_t)_{t \geq 0}$ all the pathogens of the same type are killed simultaneously. We will construct these processes so that for all $t \geq 0$ we have

$$X_t \leq Z_t.$$

At time $t = 0$ we have $X_0 = Z_0 = 1$ so that $X_0 \leq Z_0$. Assume that at some time $s > 0$ the inequality $X_s \leq Z_s$ holds. We will now show that no transition can break this inequality.

Assume first that the first transition after time s for the process $(Z_t)_{t \geq 0}$ is a birth. There are two possibilities: either the pathogen giving birth exists for $(X_t)_{t \geq 0}$ or it does not. If it does, then the birth occurs for both processes. If it does not, then the birth occurs for $(Z_t)_{t \geq 0}$ but not for $(X_t)_{t \geq 0}$. In both cases Z_t is larger than X_t after the birth.

Assume now that the first transition after time s for the process $(Z_t)_{t \geq 0}$ is a death. Again there are two possibilities: either the killed pathogen exists for $(X_t)_{t \geq 0}$ or it does not. If it does, then the pathogen is killed in both processes. Moreover, for the process $(X_t)_{t \geq 0}$ all the pathogens of the same type are killed. Hence, the death does not change the inequality between the processes. Now if the pathogen does not exist in $(X_t)_{t \geq 0}$, then it means that the inequality at time s is strict: $X_s < Z_s$ (Z_s has at least one more pathogen than X_s). The death occurs for $(Z_t)_{t \geq 0}$ but not for $(X_t)_{t \geq 0}$. Since the inequality was strict (and hence there was a difference of at least 1 between the processes) we have that $X_t \leq Z_t$ after the death.

This completes the proof that $(X_t)_{t \geq 0}$ and $(Z_t)_{t \geq 0}$ can be constructed simultaneously in a way that $X_t \leq Z_t$ for all $t \geq 0$. The simultaneous construction of two processes is called a coupling.

To complete the proof of Proposition 2.1 we need the following observation. The process $(Z_t)_{t \geq 0}$ is a binary branching process with birth rate λ and death rate 1. By Theorem 1.1 it survives if and only if $\lambda > 1$. Since we are assuming that $\lambda \leq 1$ the pathogens in $(Z_t)_{t \geq 0}$ die out with probability one. By our coupling $X_t \leq Z_t$ for all $t \geq 0$ and therefore $(X_t)_{t \geq 0}$ dies out as well. The proof of Proposition 2.1 is complete.

Our main result is the following theorem, which specifies the values of r and λ for which there is a positive probability that the pathogens survive, meaning that for all $t > 0$, there is at least one pathogen alive at time t.

Theorem 2.1. *Assume $\lambda > 0$ and r is in $[0, 1]$. The pathogens survive with positive probability if and only if $r\lambda > 1$.*

Observe that if $\lambda \leq 1$ then $r\lambda \leq r \leq 1$. Hence, survival is impossible if $\lambda \leq 1$ (which we already knew by Proposition 2.1). On the other hand, if $\lambda > 1$, then survival of pathogens (and therefore failure of the immune system) is possible if and only if $r > 1/\lambda$. This (very) simple model suggests that the immune system may indeed be overwhelmed by a virus if the virus has a mutation probability high enough.

The key to proving Theorem 2.1 is the tree of types that we now introduce.

2.1 The Tree of Types

We will construct a tree which keeps track of the genealogy of the different types of pathogens. Each vertex in the tree will be labeled by a positive integer. There will be a vertex labeled k if and only if a pathogen of type k is born at some time. We draw

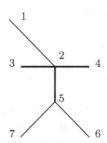

Fig. 8.1 This is an example of a tree of types. We see that type 1 has given birth to types 2 only. Type 2 has given birth to types 3, 4, and 5. Types 3 and 4 do not give birth to any new type

a directed edge from j to k if the first pathogen of type k to be born had a pathogen of type j as its parent. This construction gives a tree whose root is labeled 1 because all types of pathogens are descended from the pathogen of type 1 that is present at time zero. Since every type is eliminated eventually (why?), the pathogens survive if and only if infinitely many different types of pathogens eventually appear or, in other words, if and only if the tree described above has infinitely many vertices.

The tree of types can be thought of as representing a stochastic process $(Z_n)_{n \geq 0}$. For $n \geq 0$ we define Z_n as being the number of types in the nth generation. In Fig. 8.1 we have $Z_0 = 1$ (one type at the root of the tree), $Z_1 = 1$ (type 1 gave birth to only one type), $Z_2 = 3$ (type 2 gave birth to three types) and so on.

We will now show that the process $(Z_n)_{n \geq 0}$ is a discrete time branching process (i.e., a BGW process). Note that all the types behave independently. Knowing, for instance, that there are many type 1 pathogens does not yield any information on how type 2 is doing. More generally, once the first pathogen of type k is born, the number of mutant offspring born to type k pathogens is independent of how the other types evolve. This is so because each pathogen gives birth independently of all other pathogens and death of a pathogen affects only pathogens of the same type. Moreover, each type has the same offspring distribution: every time a new type appears it has the same offspring distribution as type 1 which initiated the process. Therefore, the tree constructed above represents a BGW tree. Hence, the process survives with positive probability if and only if the mean of the offspring distribution is strictly greater than one.

We now compute the mean offspring distribution for the tree of types.

Proposition 2.2. *The number of types that a given type gives birth to before dying has the distribution*

$$P(X = k) = \frac{(r\lambda)^k}{(r\lambda + 1)^{k+1}},$$

for $k = 0, 1, \ldots$.

Proof of Proposition 2.2. We start with the construction of the process. We attach to each pathogen three independent exponential random variables with rates 1, λr and $\lambda(1 - r)$, respectively. The rate 1 exponential corresponds to the death of the pathogen. The rate λr exponential corresponds to the birth of a pathogen with a type different from the parent. The rate $\lambda(1 - r)$ exponential corresponds to the birth of a pathogen with the same type as the parent.

We compute now the rate at which a type disappears. Whenever there are n pathogens of a given type, the type is destroyed at rate n. This is so because each of the n pathogens dies after a rate 1 exponential time, these n exponential times are independent and the type disappears when the minimum of these exponential times happens. We then use the fact that the minimum of independent exponential random variables is also an exponential and its rate is the sum of the rates.

We now turn to the rate at which a type appears. A pathogen gives birth to a pathogen of a new type after a rate $r\lambda$ exponential time. If there are n pathogens of a given type, then a pathogen of this type gives birth to a pathogen of a different type after a rate $nr\lambda$ exponential time. This is again due to the independence of the exponential times involved.

Let X be the number of types (different from type 1) that are offspring of type 1 parents. Let T be the total number of births from type 1 parents before the type disappears. We have

$$P(X = 0) = \sum_{n \geq 0} P(X = 0 | T = n) P(T = n).$$

Since each birth is independently of a different type with probability r we have

$$P(X = 0 | T = n) = (1 - r)^n.$$

Observe now that in order for $T = n$ we need n births followed by a death of a type 1 pathogen. Note also that if there are k type 1 pathogens at a given time then the rate at which a pathogen is born is $k\lambda$ and the death rate of the type is k. Hence, the probability of a birth is for any $k \geq 1$

$$\frac{k\lambda}{k\lambda + k} = \frac{\lambda}{\lambda + 1}.$$

Therefore,

$$P(T = n) = (\frac{\lambda}{\lambda + 1})^n \frac{1}{\lambda + 1}.$$

Hence,

$$P(X = 0) = \sum_{n \geq 0} (1 - r)^n (\frac{\lambda}{\lambda + 1})^n \frac{1}{\lambda + 1}.$$

By summing the geometric series we get

$$P(X = 0) = \frac{1}{1 + r\lambda}.$$

Let $k \geq 1$. We have for $n \geq k$

$$P(X = k | T = n) = \binom{n}{k} r^k (1 - r)^{n-k}.$$

For we have n births of which k are of a different type. Given that births are of a different type or not independently we get a binomial distribution. Therefore,

$$P(X = k) = \sum_{n \geq k} P(X = k | T = n) P(T = n)$$

$$= \sum_{n \geq k} \binom{n}{k} r^k (1 - r)^{n-k} \left(\frac{\lambda}{\lambda + 1}\right)^n \frac{1}{\lambda + 1}$$

Using the definition of the binomial coefficient we get

$$\binom{n}{k} = \frac{n(n - 1) \ldots (n - k + 1)}{k!}.$$

Hence,

$$P(X = k) = \left(\frac{\lambda}{\lambda + 1}\right)^k \frac{1}{\lambda + 1} \frac{r^k}{k!} \sum_{n \geq k} n(n-1) \ldots (n-k+1)(1-r)^{n-k} \left(\frac{\lambda}{\lambda + 1}\right)^{n-k}.$$

This series can be computed by the following method. For $|x| < 1$ we have

$$\frac{1}{1 - x} = \sum_{n \geq 0} x^n.$$

By differentiating both sides k times (recall that power series are infinitely differentiable and can be differentiated term by term) we get

$$\frac{k!}{(1 - x)^{k+1}} = \sum_{n \geq k} n(n - 1) \ldots (n - k + 1) x^{n-k}.$$

Hence, by letting

$$x = \frac{(1 - r)\lambda}{\lambda + 1}$$

we get

$$\sum_{n\geq k} n(n-1)\ldots(n-k+1)(1-r)^{n-k}(\frac{\lambda}{\lambda+1})^{n-k} = \frac{k!}{(1-\frac{(1-r)\lambda}{\lambda+1})^{k+1}}.$$

Therefore,

$$P(X=k) = (\frac{\lambda}{\lambda+1})^k \frac{1}{\lambda+1}\frac{r^k}{k!}\frac{k!}{(1-\frac{(1-r)\lambda}{\lambda+1})^{k+1}}.$$

A little algebra yields

$$P(X=k) = \frac{(r\lambda)^k}{(r\lambda+1)^{k+1}}.$$

This completes the proof of Proposition 2.2.

We are now ready for the proof of Theorem 2.1.

Proof of Theorem 2.1. The tree of types has a positive probability of being infinite if and only if the mean offspring $E(X) > 1$. We have

$$E(X) = \sum_{k\geq 1} kP(X=k) = \sum_{k\geq 1} k\frac{(r\lambda)^k}{(r\lambda+1)^{k+1}}.$$

Using the formula

$$\frac{1}{(1-x)^2} = \sum_{k\geq 1} kx^{k-1}$$

we get

$$E(X) = \frac{r\lambda}{(r\lambda+1)^2}\frac{1}{(1-\frac{r\lambda}{r\lambda+1})^2} = r\lambda.$$

It follows that the mean of the offspring distribution is greater than one if and only if $r\lambda > 1$. This concludes the proof of Theorem 2.1.

Problems

1. The immune response model starts with a single pathogen. What is the probability that the immune system eliminates this single pathogen before it gives birth to any other pathogen?

2. Let D be the time it takes to eliminate type 1 pathogens.

(a) Show that

$$P(D > t) \le e^{-t}.$$

(b) What is $\lim_{t \to \infty} P(D > t)$?

(c) Explain why (b) implies that type 1 pathogens will eventually be eliminated. What can we say about the other types?

3. Show that if the tree of types has finitely many vertices then the pathogens die out.

4. Let X_t be the total number of pathogens at time $t \ge 0$. We start with one pathogen so $X_0 = 1$. Let $\lambda > 0$ and $r < 1$.

(a) Explain why $(X_t)_{t \ge 0}$ is not a binary branching process.

(b) In the proof of Proposition 2.1 we introduced a continuous time binary branching process $(Z_t)_{t \ge 0}$. The only difference between $(X_t)_{t \ge 0}$ and $(Z_t)_{t \ge 0}$ is that we kill one pathogen at a time for $(Z_t)_{t \ge 0}$ instead of killing the whole type. We know that $(Z_t)_{t \ge 0}$ survives if and only if $\lambda > 1$. Why is r not relevant for the survival of $(Z_t)_{t \ge 0}$?

5. The tree of types represents a discrete time BGW with an offspring distribution given by

$$P(X = k) = \frac{(r\lambda)^k}{(r\lambda + 1)^{k+1}}$$

for $k = 0, 1, 2, \dots$.

(a) Show that the probability that this BGW survives is $1 - \frac{1}{r\lambda}$ for $r\lambda > 1$ and 0 for $r\lambda \le 1$.

(b) The tree of types is defined to analyze the process of pathogens $(X_t)_{t \ge 0}$. What does the survival probability of the tree of types represent for the process $(X_t)_{t \ge 0}$?

6. Consider the process $(X_t)_{t \ge 0}$ in the particular case when $r = 0$. We start the process with a single pathogen. Describe the evolution of the process.

7. Consider the process $(X_t)_{t \ge 0}$ in the particular case when $r = 1$. We start the process with a single pathogen.

(a) Show in this particular case $(X_t)_{t \ge 0}$ is actually a continuous time binary branching process. (Recall that no two types are the same.)

(b) Give a direct proof (not using the results of this section) that the pathogens survive if and only if $\lambda > 1$.

8. Let $k \geq 1$ and x in $(-1, 1)$.

(a) Show (by induction) that the kth derivative of $\frac{1}{1-x}$ is

$$\frac{k!}{(1-x)^{k+1}}.$$

(b) Show that the kth derivative of $\sum_{n \geq 0} x^n$ is

$$\sum_{n \geq k} n(n-1) \ldots (n-k+1)x^{n-k}.$$

(c) Show that

$$\frac{k!}{(1-x)^{k+1}} = \sum_{n \geq k} n(n-1) \ldots (n-k+1)x^{n-k}.$$

3   A Model for Virus Survival

Compared to other species a virus replicating through (single strand) RNA has a very high mutation rate and a great deal of genomic diversity. From the virus point of view a high mutation rate is advantageous because it may create rather diverse virus genomes, this may overwhelm the immune system of the host and ensure survival of the virus population. This was seen in the previous section. On the other hand, a high mutation rate may result in many nonviable individuals and hurt survival. It seems therefore that mutation rates should be high but not too high in order for the virus to survive. This is the hypothesis we will test in this section. In fact, we will show that our model allows survival of the virus for even very high mutation rate. This contradicts the widely accepted hypothesis that survival should not be possible above a certain mutation threshold.

Note that the immune response model of the previous section gives the same birth rate to all types. In this section each type will have a different birth rate. This is necessary in order to test the hypothesis of this section.

We now describe our model. We have two parameters: $a > 0$ and r in $[0, 1]$. Start with one individual at time 0, and sample a birth rate λ from the uniform distribution on $[0, a]$. Recall that the uniform distribution on $[0, a]$ has a flat density

$$f(x) = \frac{1}{a} \text{ for all } x \in [0, a],$$

and that its expected value is

$$E(X) = \frac{a}{2}.$$

The initial individual gives birth at rate λ and dies at rate 1. As always these are the rates of exponential distributions. Every time there is a birth the new individual: (1) with probability $1 - r$ keeps the same birth rate λ as its parent, and (2) with probability r is given a new birth rate λ', sampled independently of everything else from the same uniform distribution. We think of r as the mutation probability and the birth rate of an individual as representing the fitness or genotype of the individual. Since uniform distributions are continuous, a genotype cannot appear more than once (why?). For convenience we label the genotypes (or types) in the order of their appearance.

We say that the virus survives if the probability that there is at least one virus at all times is positive. Hence, the virus dies out if after a finite (random) time no virus is left. The first question we address is whether the initial type 1 virus may survive forever. More generally, we now give a necessary and sufficient condition for a fixed type to survive.

Proposition 3.1. *A fixed type survives if and only if $a(1 - r) > 1$.*

Intuitively, a large a allows for large λ's to appear and therefore should help the virus to survive. However, even if a is large Proposition 3.2 shows that a fixed type can survive only if the mutation probability r is not too large (why?).

Proof of Proposition 3.1. All types appear through a single individual whose λ has been sampled from the uniform distribution on $[0, a]$. Hence, the survival probability is the same for any given type. For the sake of concreteness we concentrate on type 1. Let X_t be the number of type 1 individuals alive at time t. A type 1 individual gives birth at rate λ and dies at rate 1. The new individual is a type 1 individual with probability $1 - r$ and of a different type with probability r. Hence, a type 1 individual gives birth to another type 1 individual at rate $\lambda(1 - r)$. Hence, conditional on the rate λ, X_t is a continuous time binary branching process.

Let A be the event that X_t survives. That is,

$$A = \{X_t > 0, \quad \forall t > 0\}.$$

By Theorem 1.1, $P(A|\lambda) > 0$ if and only if $\lambda(1 - r) > 1$. By conditioning on λ,

$$P(A) = \frac{1}{a} \int_0^a P(A|\lambda) d\lambda.$$

If $a(1 - r) \leq 1$, then $\lambda(1 - r) < 1$ and $P(A|\lambda) = 0$ for every λ in $[0, a]$. Hence,

$$P(A) = \frac{1}{a} \int_0^a P(A|\lambda) d\lambda = 0.$$

On the other hand, if $a(1-r) > 1$, then $P(A|\lambda) > 0$ for λ in $[\frac{1}{1-r}, a]$. It turns out that this is enough to show that

$$\frac{1}{a} \int_0^a P(A|\lambda) d\lambda > 0 \tag{3.1}$$

and therefore that $P(A) > 0$. We will indicate the proof of (3.1) in the Problems. This completes the proof of Proposition 3.1.

We now turn to the second (and more interesting) way the virus may survive.

First, we need the tree of types that was introduced for another model in the preceding section. We recall the definition. Each vertex in the tree will be labeled by a positive integer. There will be a vertex labeled k if and only if a virus of type k is born at some time. We draw a directed edge from j to k if the first virus of type k to be born had a virus of type j as its parent. Similarly to the model in Sect. 2 each type has the same offspring (i.e., the number of types that each type gives birth to) distribution and each type is independent of all other types. Hence, here too the tree represents a discrete time BGW process.

With the tree of types in hand we are ready to give necessary and sufficient conditions for the survival of the virus population.

Since the tree of types is a BGW it is infinite if and only if the corresponding mean offspring distribution is strictly larger than 1. We now proceed to compute this mean. Let $m(r)$ be the mean number of types that are born from type 1 individuals.

Proposition 3.2. *Let $m(r)$ be the mean number of types that are born from a fixed type. If $a(1-r) < 1$, then*

$$m(r) = \frac{1}{a} \int_0^a \frac{r\lambda}{1 - (1-r)\lambda} d\lambda.$$

If $a(1-r) \geq 1$, then $m(r) = +\infty$.

Proof of Proposition 3.2. Let X_t be the number of type 1 individuals alive at time t. Let Y_t be the number of individuals born up to time t that are offspring of genotype 1 individuals and have a different type. That is, Y_t counts all the new types born from type 1 parents up to time t. Some types will have disappeared by time t but are still counted.

Let $h > 0$ be close to 0. Each type 1 virus gives birth between times t and $t + h$ to a single new type with probability λrh. The probability that a type 1 virus gives birth to 2 or more new types is of order h^2 or higher for small h. Therefore,

$$E(Y_{t+h} - Y_t|\lambda) = \lambda rh E(X(t)) + o(h).$$

By dividing both sides by h and letting h go to 0 it follows that

$$\frac{d}{dt}E(Y_t|\lambda) = \lambda r E(X_t|\lambda).$$

Given λ, X_t is a continuous time BGW and so by Theorem 1.1

$$E(X_t|\lambda) = \exp((\lambda(1-r)-1)t).$$

Hence,

$$\frac{d}{dt}E(Y_t|\lambda) = \lambda r \exp((\lambda(1-r)-1)t).$$

Integrating between times 0 and t yields

$$E(Y_t|\lambda) = r\lambda \int_0^t \exp((\lambda(1-r)-1)s)ds = \frac{r\lambda}{-1+(1-r)\lambda}[\exp((\lambda(1-r)-1)t)-1].$$

Conditioning with respect to λ (which is uniformly distributed on $[0,a]$)

$$E(Y_t) = \frac{1}{a}\int_0^a E(Y_t|\lambda)d\lambda = \frac{1}{a}\int_0^a \frac{r\lambda}{-1+(1-r)\lambda}[\exp((\lambda(1-r)-1)t)-1]d\lambda.$$

By definition, $E(Y_t)$ is the expected number of types that type 1 individuals give birth to up to time t. Hence, the limit of $E(Y_t)$ as t goes to infinity is the expected total number of types that are ever born to type 1 individuals. That is, this limit is $m(r)$. We now compute this limit.

$$\lim_{t\to\infty} E(Y_t) = \frac{1}{a}\int_0^a \frac{r\lambda}{-1+(1-r)\lambda}\lim_{t\to\infty}[\exp((\lambda(1-r)-1)t)-1]d\lambda.$$

Interchanging the limit and the integral as we did above is not always possible. Here, it can be justified but we omit this part.

At this point there are two possibilities.

- If $a(1-r) \le 1$ we have that $\lambda(1-r)-1 < 0$ for all λ in $[0,a)$. Hence,

$$\lim_{t\to\infty} \exp((\lambda(1-r)-1)t) = 0.$$

Therefore,

$$m(r) = \lim_{t\to\infty} E(Y_t) = \frac{1}{a}\int_0^a \frac{r\lambda}{1-(1-r)\lambda}d\lambda.$$

Note that if $a(1-r) < 1$ this is a proper integral. On the other hand, if $a(1-r) = 1$ this is an improper integral (singularity at $\lambda = a$) that diverges (why?). Hence, $m(r) = +\infty$ for $a(1-r) = 1$.

- If $a(1-r) > 1$, then for λ in $(\frac{1}{1-r}, a]$ we have

$$\lim_{t\to\infty} \exp((\lambda(1-r) - 1)t) = +\infty$$

and

$$m(r) = \lim_{t\to\infty} E(Y_t) = +\infty.$$

This completes the proof of Proposition 3.2.

We are now ready for the main result of this section.

Proposition 3.3. *The virus survives if and only the tree of types is infinite. That is, if and only if $m(r) > 1$.*

Proof of Proposition 3.3. One direction is easy. If the tree of types is infinite it means that the process gave birth to infinitely many individuals and hence survives forever.

For the converse assume that the tree of types is finite. Since this is a BGW we must have $m(r) \le 1$. By Proposition 3.2 we have $a(1-r) < 1$. Therefore, by Proposition 3.1 any fixed type dies out. Hence, only finitely many types appeared and each one of these types died out after a finite random time. Since there are only finitely many types all types will have disappeared after a finite random time (why?). Therefore, the process dies out after a finite random time. This completes the proof of Proposition 3.3.

We now turn to explicit computations.

Proposition 3.4. *Let $a > 2$. Then the virus survives if the mutation probability r is large enough.*

Proposition 3.4 shows that the hypothesis that the virus is doomed if the mutation probability is too high is not true when $a > 2$.

Proof of Proposition 3.4. It is enough to show that the tree of types is infinite. Recall that the mean offspring for the tree is

$$m(r) = \frac{1}{a} \int_0^a \frac{r\lambda}{1 - (1-r)\lambda} d\lambda.$$

Since $(1-r)\lambda \ge 0$ we have $1 - (1-r)\lambda \le 1$ and

$$m(r) = \frac{1}{a} \int_0^a \frac{r\lambda}{1 - (1-r)\lambda} d\lambda \ge \frac{1}{a} \int_0^a r\lambda d\lambda = \frac{1}{2} ra.$$

Therefore, if $r > \frac{2}{a}$ then $m(r) > 1$ and the tree of types is infinite. Note that because we picked $a > 2$ we have that $\frac{2}{a} < 1$. The tree of types is infinite for any r in $(\frac{2}{a}, 1]$. This completes the proof of Proposition 3.4.

It turns out that $a = 2$ is a critical value. We will show in the Problems that if $1 < a < 2$ then survival is not possible when r is too large. Therefore, the model behaves quite differently depending on whether $a < 2$ or not.

The last case we deal with is $a \le 1$.

Proposition 3.5. *Assume that $0 < a \le 1$. Then survival is not possible for any r in* $[0, 1]$.

Proposition 3.5 is not really surprising. Since $\lambda \le a$ the birth rate is always less than the death rate (which is 1) and we expect the virus to die out.

Proof of Proposition 3.5. Recall that

$$m(r) = \frac{1}{a} \int_0^a \frac{r\lambda}{1 - (1 - r)\lambda} d\lambda.$$

We compute the following derivative

$$\frac{d}{dr} \frac{r\lambda}{1 - (1 - r)\lambda} = \frac{\lambda(1 - \lambda)}{(r\lambda + 1 - \lambda)^2}.$$

Observe that for λ in $[0, a]$ and $a \le 1$ this derivative is positive. Hence,

$$\frac{r\lambda}{1 - (1 - r)\lambda}$$

is increasing as a function of r. Therefore, $m(r)$ is also increasing as a function of r (why?). This implies that for all r in $[0, 1]$

$$m(r) \le m(1) = \frac{1}{a} \int_0^a \lambda d\lambda = \frac{a}{2} \le \frac{1}{2}.$$

That is, $m(r) < 1$ for all r. Hence, the tree of types is finite. By Proposition 3.3 the process dies out. This completes the proof of Proposition 3.5.

Remark. By doing a finer analysis it is possible to have a complete picture of the behavior of the model. In particular, for $1 < a < 2$ there exists a critical value for r above which survival is possible and below which it is not. For $a > 2$, survival is possible for all r in $[0, 1]$. For $a = 2$ survival is possible for all r except $r = 1$. See the notes and references at the end of the chapter.

Problems

1. In the model for immune response of Sect. 2 any fixed type dies out. On the other hand, for the virus population model of this section we know a fixed type may survive by Proposition 3.2. Explain what makes the models behave differently.

2. Show that the same type cannot appear twice.

3. Let $a > 1$. Show that a fixed type can survive if and only $r < 1 - 1/a$.

4. Show that for $a(1 - r) < 1$

$$m(r) = -\frac{r}{1-r} - \frac{1}{a}\frac{r}{(1-r)^2} \ln(1 - a(1 - r)).$$

5. Use the expression for $m(r)$ found in Problem 4 to plot m as a function of r (for $r > 1 - 1/a$) when

(a) $a = 5/4$.
(b) $a = 7/4$.
(c) $a = 3$.
(d) Intuitively, the larger r the more types should appear and therefore the larger $m(r)$ should be. Is this intuition correct?

6. Proposition 3.4 shows that when $a > 2$ the virus survives provided r is large enough. In this problem we show that when $1 < a < 2$ the virus dies out when r is too large. Hence, the behavior of the model changes drastically depending on the value of a.

(a) Show that if λ is in $[0, a]$ then

$$1 - (1 - r)\lambda \geq 1 - (1 - r)a.$$

(b) Use (a) to show that

$$m(r) = \frac{1}{a} \int_0^a \frac{r\lambda}{1 - (1 - r)\lambda} d\lambda \leq \frac{1}{a} \int_0^a \frac{r\lambda}{1 - (1 - r)a} d\lambda = \frac{1}{2}\frac{ar}{1 - (1 - r)a}.$$

(c) Use (b) to show that $m(r) < 1$ when r is above a certain threshold.
(d) Show that the virus die out when r is large enough.
(d) Compute a numerical value for r above which the virus die out when $a = 3/2$.

7. Assume that $r = 0$. That is, there is no mutation. Show that survival is possible if and only if $a > 1$.

8. Assume that $r = 1$. That is, every birth has a different type.

(a) Assume that type 1 has birth rate λ. Let X be the number of offspring of the type 1 individual. Show that for all $k \geq 0$ we have

$$P(X = k \,|\, \lambda) = (\frac{\lambda}{\lambda + 1})^k \frac{1}{\lambda + 1}.$$

(b) Show that

$$E(X \,|\, \lambda) = \lambda.$$

(c) Show that

$$E(X) = \frac{a}{2}.$$

(d) Show that survival is possible if and only if $a > 2$.

9. Consider the process $(Z_t)_{t \geq 0}$ for which an individual gives birth to a single particle after an exponential time with rate λ_1 or dies after an exponential time with rate 1. All individuals evolve independently of each other. Consider $(X_t)_{t \geq 0}$ which evolves as $(Z_t)_{t \geq 0}$ except that its birth rate is λ_2. Assume also that $\lambda_2 > \lambda_1$. Start both processes with a single individual. Construct simultaneously $(Z_t)_{t \geq 0}$ and $(X_t)_{t \geq 0}$ so that for all $t \geq 0$ we have

$$Z_t \leq X_t.$$

(A similar construction is given in the proof of Proposition 2.1).

10. Consider a function f strictly positive and increasing on $[c, d]$ for $c < d$. Show that

$$\int_c^d f(t) dt > 0.$$

11. Use problems 9 and 10 to prove (3.1). (Let $f(\lambda)$ be $P(A \,|\, \lambda)$).

4 A Model for Bacterial Persistence

Since at least Bigger (1944) it is known that antibiotic treatment will not completely kill off a bacteria population. For many species a small fraction of bacteria is not sensitive to antibiotics. These bacteria are said to be "persistent." It turns out that persistence is not a permanent state. In fact a bacterium can switch back and forth between a "normal" (i.e. non-persistent) state and a "persistent" state. In a normal state a bacterium can reproduce but is killed by an antibiotic attack. In the persistent state a bacterium does not reproduce (or very seldom) but is resistant to antibiotics.

We propose a model in this section that will show that persistence is a very good strategy (from the bacterial point of view) in the sense that even under stressful conditions it can ensure survival of the bacteria.

We now describe the model. Consider the following continuous time model. Bacteria can be in one of two states. We think of state 1 as being the normal (i.e. non-persistent) state and state 2 as being the persistent state. Note that the normal state is vastly predominant in the population. A bacterium in state 1 is subject to two possible transitions. It can give birth at rate λ to another state 1 individual or it can switch to state 2 at rate a. A bacterium in state 2 has only one possible transition. It can switch to state 1 at rate b. Moreover, the bacteria in state 1 can be killed in the following way. Let $T_i = iT$ for $i \geq 1$ where T is a fixed positive constant. This defines a sequence of killing times $T_1, T_2, \ldots$. At each killing time T_i all the bacteria in state 1 are killed but the bacteria in state 2 are unaffected.

The model mimics the observed behavior. State 1 bacteria multiply until disaster strikes and then they all die. State 2 bacteria cannot give birth but persist under disasters. Hence, state 2 bacteria ensure survival through disasters but cannot give birth.

The main question we are concerned with is for which parameter values do the bacteria survive? The following result answers this question.

Theorem 4.1. *For any $a > 0$, $b > 0$ and $\lambda > 0$ there is a critical value T_c such that the bacteria survive forever with positive probability if and only if $T > T_c$.*

Note that if at some killing time T_i all the bacteria are in state 1 (which is possible) then the bacteria die out. So it is not so clear whether this model allows for survival. Theorem 4.1 shows that it does.

The exact value of T_c depends on the parameters a, b, and λ. However, T_c varies very little with a and b, see the problems. This shows that survival is possible for a wide range of parameters.

Proof of Theorem 4.1. We start the model with finitely many bacteria. We define an auxiliary discrete time stochastic process Z_n, $n \geq 0$. We wait until the first killing time T and we let Z_0 be the number of bacteria in state 2 at time T. If $Z_0 = 0$, we set $Z_i = 0$ for $i \geq 1$. If $Z_0 \geq 1$, then we wait until the second killing time $2T$ and let Z_1 be the number of state 2 bacteria at time $2T$. More generally, for any $k \geq 1$ let Z_k be the number of state 2 individuals at the $(k + 1)$th killing time $(k + 1)T$.

We claim that the process $(Z_k)_{k \geq 0}$ is a discrete time BGW. This can be seen using the following argument. Each state 2 bacterium present at time $T_1 = T$ starts a patch of bacteria. Note that in order for the patch to get started we first need the initial state 2 bacterium to switch to state 1. At the second killing time T_2 all the individuals in state 1 are killed and we are left with Z_1 individuals in state 2. If $Z_1 = 0$, the bacteria have died out. If $Z_1 \geq 1$, then each 2 present at time T_1 starts its own patch. Each patch will live between times T_2 and T_3. These patches are independent and identically distributed. At time T_3 each 2 (if any) starts a new patch and so on. This construction shows that $(Z_k)_{k \geq 0}$ is a discrete time BGW. Moreover, the bacteria population survives forever if and only if $Z_k \geq 1$ for all

$k \geq 0$. For if $Z_k = 0$ for some $k \geq 1$ this means at the corresponding killing time T_{k+1} there are no type 2 individuals and no type 1 individuals either. That is, the bacteria have died out. If, on the other hand, $Z_k \geq 1$ for all $k \geq 0$, the bacteria will survive forever.

Now, the process $(Z_k)_{k \geq 0}$ survives if and only if $E(Z_1 | Z_0 = 1) > 1$. Hence, the problem of survival for the bacteria population is reduced to computing the expected value $E(Z_1 | Z_0 = 1)$. We do this computation next.

For $t < T_1$ let $x(t)$ be the expected number of type 1 bacteria and $y(t)$ be the expected number of type 2 bacteria at time t, starting at time 0 with a single type 2 and no type 1. We have for $h > 0$

$$x(t + h) - x(t) = \lambda h x(t) - a h x(t) + b h y(t) + o(h),$$

$$y(t + h) - y(t) = a h x(t) - b h y(t) + o(h).$$

This is so because in the time interval $[t, t + h]$ a type 1 bacterium gives birth to another type 1 bacterium at rate λh or switches to a type 2 at rate ah. On the other hand, a type 2 bacterium switches to a type 1 at rate bh.

By dividing both equations by h and letting h to 0 we get the following system of differential equations

$$\frac{d}{dt}x(t) = (\lambda - a)x(t) + by(t),$$

$$\frac{d}{dt}y(t) = ax(t) - by(t).$$

This is a linear system with constant coefficients. The corresponding matrix

$$A = \begin{pmatrix} \lambda - a & b \\ a & -b \end{pmatrix}$$

has two real distinct eigenvalues v_1 and v_2

$$v_1 = \frac{-a - b + \lambda + \sqrt{\Delta}}{2}$$

$$v_2 = \frac{-a - b + \lambda - \sqrt{\Delta}}{2},$$

where

$$\Delta = (a + b - \lambda)^2 + 4b\lambda > (a + b - \lambda)^2.$$

Note that the determinant of A is $-\lambda b < 0$. The determinant is also the product of the two eigenvalues. Hence, the eigenvalues have opposite signs. Since $v_1 > v_2$ we must have $v_1 > 0$ and $v_2 < 0$. A standard computation yields the solution of the system of differential equations, see, for instance, Hirsch and Smale (1974). We get

$$x(t) = c_2 \frac{b - a + \lambda + \sqrt{\Delta}}{2a} \exp(v_1 t) - c_1 \frac{a - b - \lambda + \sqrt{\Delta}}{2a} \exp(v_2 t)$$

$$y(t) = c_1 \exp(v_2 t) + c_2 \exp(v_1 t).$$

We can find the constants c_1 and c_2 by using the initial conditions $x(0) = 0$ and $y(0) = 1$. We get

$$c_1 = \frac{b - a + \lambda + \sqrt{\Delta}}{2\sqrt{\Delta}} \quad \text{and} \quad c_2 = 1 - c_1.$$

Since $\sqrt{\Delta} > |a + b - \lambda|$ we have $c_1 > 0$. It is also possible to check that $c_1 < 1$. Hence, $c_2 > 0$.

The function y drops from $y(0) = 1$ to some minimum and then increases to infinity as t goes to infinity. Hence, there is a unique $T_c > 0$ such that $y(T_c) = 1$. We will check these claims on a particular case in the problems section.

The critical value T_c can be computed numerically by solving the equation $y(t) = 1$. For any $T > T_c$ the Galton–Watson process Z_k is super-critical and survives forever with a positive probability. For $T \leq T_c$ the process Z_k dies out. This completes the proof of Theorem 4.1.

Problems

In all the problems below we set $\lambda = 2$, $a = b = 1$.

1. Show that the eigenvalues of A are $v_1 = \sqrt{2}$ and $v_1 = -\sqrt{2}$.

2. Check that the solution of the system of differential equations

$$\frac{d}{dt} x(t) = x(t) + y(t),$$

$$\frac{d}{dt} y(t) = x(t) - y(t).$$

with $x(0) = 0$ and $y(0) = 1$
is

$$x(t) = \frac{\sqrt{2}}{4} e^{\sqrt{2}t} - \frac{\sqrt{2}}{4} e^{-\sqrt{2}t},$$

$$y(t) = \frac{2 - \sqrt{2}}{4} e^{\sqrt{2}t} + \frac{2 + \sqrt{2}}{4} e^{-\sqrt{2}t}.$$

3. Show that the function y defined in Problem 2 has a minimum at

$$t_0 = \frac{\sqrt{2}}{2} \ln(\sqrt{2} + 1).$$

4. Recall that $y(t)$ is the expected number of bacteria in state 2 at time t. Did you expect the function y to behave as described in Problem 3? Why or why not?

5. (a) Let $X = e^{\sqrt{2}t}$ in the equation $y(t) = 1$ and show that the equation can be transformed into

$$X^2 - 2\sqrt{2}(\sqrt{2} + 1)X + (\sqrt{2} + 1)^2 = 0.$$

(b) Show that $X = 1$ is a solution of the equation (a).
(c) Show that the other solution of the equation in (a) is $X = (\sqrt{2} + 1)^2$.
(d) Use (c) to show that the solutions of $y(t) = 1$ are $t = 0$ and

$$T_c = \sqrt{2}\ln(1 + \sqrt{2}).$$

6. Graph the critical value T_c as a function of a and b. Let the parameters a and b vary from 10^{-6} to 10^{-3} and set λ equal to 2. Interpret the graph.

Notes

Section 2 is based on Schinazi and Schweinsberg (2008) where the reader can find several other models for immune response. Section 3 is based on Cox and Schinazi (2012). In that reference a complete analysis of the virus model is done. Section 4 comes from Garet et al. (2012) where a model with random killing times is also treated.

References

Bigger, J.W.: Treatment of staphylococcal infections with penicillin. Lancet **244**, 497–500 (1944)
Cox, J.T., Schinazi, R.B.: A branching process for virus survival. J. Appl. Probab. **49**, 888–894 (2012)
Garet, O., Marchand, R., Schinazi, R.B.: Bacterial persistence: a winning strategy? Markov Process. Relat. Fields **18**, 639–650 (2012)
Hirsch, M.W., Smale, S.: Differential equations, dynamical systems and linear algebra. Academic Press (1974)
Schinazi, R.B., Schweinsberg, J.: Spatial and non spatial stochastic models for immune response. Markov Process. Relat. Fields **14**, 255–276 (2008)

Chapter 9
Continuous Time Birth and Death Chains

In this chapter we look at a more general class of continuous time Markov processes. The main constraint is that the process changes state by one unit at a time. An important tool to study these processes is again differential equations.

1 The Kolmogorov Differential Equations

A continuous time birth and death chain is a stochastic process that we denote by $(X_t)_{t\geq0}$. For all $t \geq 0$ X_t is a positive integer or 0. This process jumps at random times by one unit at a time. More precisely, let $(\lambda_i)_{i\geq0}$ and $(\mu_i)_{i\geq0}$ be two sequences of positive real numbers. If $X_t = i \geq 1$, then after an exponential time with rate $\lambda_i + \mu_i$ the chain jumps to $i + 1$ with probability $\frac{\lambda_i}{\lambda_i+\mu_i}$ or jumps to $i - 1$ with probability $\frac{\mu_i}{\lambda_i+\mu_i}$. We take $\mu_0 = 0$. That is, the transition from 0 to -1 is forbidden. The chain remains on the positive integers at all times. If $X_t = 0$ after an exponential time with rate λ_0, the chain jumps to 1. The rates λ_i and μ_i are called the birth and death rates of the chain, respectively.

We now give several examples of birth and death chains.

Example 1.1. Consider a continuous time random walk on the half line defined as follows. After a mean 1 exponential time the walker jumps to the right with probability p or to the left with probability q. Show that this is a birth and death chain.

We are going to find the rates λ_i and μ_i for this chain. Let $\lambda_i = p$ for $i \geq 1$, $\lambda_0 = 1$, $\mu_i = q$ for $i \geq 1$ and $\mu_0 = 0$. Then after a random time with parameter $\lambda_i + \mu_i = p + q = 1$ the walker moves to $i + 1$ with probability $\frac{\lambda_i}{\lambda_i+\mu_i} = p$ or moves to $i-1$ with probability $\frac{\mu_i}{\lambda_i+\mu_i} = q$. This shows that this is a continuous birth and death chain.

© Springer Science+Business Media New York 2014
R.B. Schinazi, *Classical and Spatial Stochastic Processes: With Applications to Biology*, DOI 10.1007/978-1-4939-1869-0_9

Example 1.2. Consider a bank with N tellers. Customers arrive at rate λ. Service of a customer is also exponentially distributed with rate μ. Let X_t be the number of customers being helped or waiting in line. Show that X_t is a birth and death chain.

Assume $X_t = i \geq 0$. A new customer arrives at rate λ. Hence, we let $\lambda_i = \lambda$ for $i \geq 0$. Let $i \geq 1$. If $i \leq N$, then all the customers in the system are being helped. Each service happens at rate μ. Therefore, X_t decreases by 1 when the minimum of i independent exponential random variables happens. This minimum is also an exponential random variable and it has rate $i\mu$ (see the properties of the exponential distribution). So we let $\mu_i = i\mu$ for $1 \leq i \leq N$. If $i > N$, then $\mu_i = N\mu$. This is so because we have N tellers and therefore at any given time at most N customers can be helped.

Example 1.3. A Poisson process is a birth and death chain.

If the birth rate of the Poisson process is $\lambda > 0$, then we set $\lambda_i = \lambda$ and $\mu_i = 0$ for all integers $i \geq 0$. Hence, the Poisson process is a birth and death chain.

Example 1.4. Some continuous time branching processes are also birth and death chains. Consider a population where each individual gives birth at rate λ or dies at rate μ. Assume also all individuals behave independently. We now show that the number of individuals alive at time t is a birth and death chain.

Assume that there are n individuals at some time t. Then there is a birth in the population at rate $n\lambda$ and a death at rate $n\mu$. That is, this is a birth and death chain with rates $\lambda_n = n\lambda$ and $\mu_n = n\mu$ for all $n \geq 0$.

The birth and death process $(X_t)_{t \geq 0}$ has the following Markov property. For all times s and t, for all positive integers, i, j, and for any collection of positive integers $(k_u)_{0 \leq u < s}$ we have

$$P(X_{t+s} = j | X_s = j; X_u = k_u \text{ for } 0 \leq u < s) = P(X_{t+s} = j | X_s = j).$$

In words, the Markov property states that the process depends on the past and the present only through the present. The process $(X_t)_{t \geq 0}$ also has the following homogeneous property. For any integers i, j and any real numbers $t \geq 0$ and $s \geq 0$ we have

$$P(X_{t+s} = j | X_s = i) = P(X_t = j | X_0 = i).$$

Let $i \geq 0$ and $j \geq 0$ be two integers. We define the transition probability $P_t(i, j)$ by

$$P_t(i, j) = P(X_t = j | X(0) = i).$$

That is, $P_t(i, j)$ is the probability that the chain is in state j at time t given that it started at time 0 in state i. We now turn to the properties of the transition probabilities.

In most cases we cannot find $P_t(i, j)$. However, they are the solution of certain differential equations and this turns out to be quite helpful.

Proposition 1.1. *The transition probabilities of a birth and death chain satisfy a system of differential equations known as the backward Kolmogorov differential equations. These are given by*

$$\frac{d}{dt} P_t(i, j) = \mu_i P_t(i-1, j) - (\lambda_i + \mu_i) P_t(i, j) + \lambda_i P_t(i+1, j) \text{ for } i \geq 1 \text{ and } j \geq 0$$

and

$$\frac{d}{dt} P_t(0, j) = -\lambda_0 P_t(0, j) + \lambda_0 P_t(1, j) \text{ for } j \geq 0$$

together with the initial condition $P_0(i, j) = 0$ for $i \neq j$ and $P_0(i, i) = 1$.

Sketch of the Proof of Proposition 1.1. We will give an informal derivation of this system of differential equations.

Assume that $i \geq 1$ and $j \geq 0$. Conditioning on what happens between times 0 and h we get the following.

$$P_{t+h}(i, j) = h\lambda_i P_t(i+1, j) + h\mu_i P_t(i-1, j) + (1 - (\lambda_i + \mu_i)h) P_t(i, j) + o(h) \tag{1.1}$$

This equation is obtained as follows. Assume that there is a birth between times 0 and h. Since the chain is at i this happens with rate λ_i. The chain jumps to $i + 1$. Hence, the chain now has to go from $i + 1$ to j between times h and $t + h$. Because of the Markov property (we only need the state of the chain at time h to have the distribution of the chain at further times) and the homogeneous property of the process (the distribution of the chain between times h and $t + h$ is the same as the distribution between times 0 and t) this has probability $P_t(i + 1, j)$. This explains the first term on the r.h.s of (1.1). The second term on the r.h.s. of (1.1) corresponds to a death between times 0 and h and is obtained in a similar way as the first term. The third term on the r.h.s. of (1.1) corresponds to nothing happening between times 0 and h. Finally, $o(h)$ corresponds to the probability that two or more events happen between times 0 and h. In fact, the probability that two or more events happen between times 0 and h is less than h^2 (which is an $o(h)$).

Going back to (1.1) we get

$$\frac{P_{t+h}(i, j) - P_t(i, j)}{h} = \lambda_i P_t(i+1, j) + \mu_i P_t(i-1, j) - (\lambda_i + \mu_i) P_t(i, j) + \frac{o(h)}{h}.$$

Letting h go to 0 we get

$$\frac{d}{dt} P_t(i, j) = \mu_i P_t(i - 1, j) - (\lambda_i + \mu_i) P_t(i, j) + \lambda_i P_t(i + 1, j).$$

This is the backward Kolmogorov equation when $i \geq 1$. Note that if $i = 0$ then there can be no death and this actually simplifies the equations. The case $i = 0$ is left as an exercise. This completes the proof of Proposition 1.1.

Observe that we have found the backward equations by conditioning on what happens between times 0 and h. We get another set of equations if we condition on what happens between times t and $t + h$. These are the so-called forward Kolmogorov differential equations. We give an informal derivation of the forward equations. Let $i \geq 0$ and $j \geq 1$ be two states. We have

$$P_{t+h}(i, j) = \lambda_{j-1} h P_t(i, j-1) + \mu_{j+1} h P_t(i, j+1) + (1 - (\lambda_i + \mu_i)h) P_t(i, j) + o(h).$$

Thus,

$$\frac{P_{t+h}(i, j) - P_t(i, j)}{h} = \lambda_{j-1} P_t(i, j-1) - (\lambda_j + \mu_j) P_t(i, j) + \mu_{j+1} P_t(i, j+1) + \frac{o(h)}{h}.$$

Letting h go to 0 yields the forward Kolmogorov equations

$$\frac{d}{dt} P_t(i, j) = \lambda_{j-1} P_t(i, j-1) - (\lambda_j + \mu_j) P_t(i, j) + \mu_{j+1} P_t(i, j+1).$$

So we have proved

Proposition 1.2. *The transition probabilities of a birth and death chain satisfy a system of differential equations known as the forward Kolmogorov differential equations:*

$$\frac{d}{dt} P_t(i, j) = \lambda_{j-1} P_t(i, j-1) - (\lambda_j + \mu_j) P_t(i, j) + \mu_{j+1} P_t(i, j+1) \text{ for } i \geq 0 \text{ and } j \geq 1$$

and for $j = 0$ we get

$$\frac{d}{dt} P_t(i, 0) = -\lambda_0 P_t(i, 0) + \mu_1 P_t(i, 1) \text{ for } i \geq 0.$$

In very few cases we can integrate the Kolmogorov differential equations and get an explicit expression for $P_t(i, j)$. Even so these differential equations are a valuable tool in the analysis of continuous time stochastic processes.

1.1 The Pure Birth Processes

We consider a pure birth process. That is, $\mu_i = 0$ for all i. The forward equations simplify to

$$\frac{d}{dt} P_t(i, j) = \lambda_{j-1} P_t(i, j-1) - \lambda_j P_t(i, j) \text{ for } i \geq 0 \text{ and } j \geq 1.$$

This is a first order system of linear equations. Recall the following.

Lemma 1.1. *Assume that g is continuous and that λ is a constant. The first order linear equation*

$$y' = -\lambda y + g$$

has a unique solution given by

$$y(t) = y(0)e^{-\lambda t} + e^{-\lambda t} \int_0^t e^{\lambda s} g(s) ds.$$

The Lemma can be easily proved by checking that y is indeed a solution of the differential equation. See the exercises.

Applying Lemma 1.1 to the forward equation with $g(t) = \lambda_{j-1} P_t(i, j-1)$ we get

$$P_t(i, j) = P_0(i, j)e^{-\lambda_j t} + e^{-\lambda_j t} \int_0^t e^{\lambda_j s} \lambda_{j-1} P_s(i, j-1) ds \tag{1.2}$$

Since there are no deaths, the chain moves only to the right and $P_t(i, j) = 0$ for $j < i$. For the same reason, the only way the chain can go from i to i is to stay at i. That is, the exponential time with rate λ_i must be larger than t. Hence,

$$P_t(i, i) = \int_t^{+\infty} \lambda_i e^{-\lambda_i s} ds = e^{-\lambda_i t}.$$

Assume now that $j > i$, then $P_0(i, j) = 0$ and (1.2) yields

$$P_t(i, j) = e^{-\lambda_j t} \int_0^t e^{\lambda_j s} \lambda_{j-1} P_s(i, j-1) ds \tag{1.3}$$

If we let $j = i + 1$, we get

$$P_t(i, i+1) = e^{-\lambda_{i+1} t} \int_0^t e^{\lambda_{i+1} s} \lambda_i e^{-\lambda_i s} ds.$$

Thus,

$$P_t(i, i+1) = \frac{\lambda_i}{\lambda_{i+1} - \lambda_i} (e^{-\lambda_i t} - e^{-\lambda_{i+1} t}) \text{ if } \lambda_i \neq \lambda_{i+1}$$

and

$$P_t(i, i+1) = \lambda_i t e^{-\lambda_i t} \text{ if } \lambda_i = \lambda_{i+1}.$$

We may find any $P_t(i, j)$ recursively as follows. Using (1.2) for $j = i + 2$ we get

$$P_t(i, i+2) = e^{-\lambda_j t} \int_0^t e^{\lambda_j s} \lambda_{i+1} P_s(i, i+1) ds.$$

Since $P_s(i, i+1)$ is known we may compute the integral and have an explicit formula for $P_t(i, i+2)$. We may iterate this method to get $P_t(i, i+k)$ for all $k \geq 0$.

1.2   The Yule Process

This is a particular case of pure birth. Let $\lambda_i = i\lambda$ and $\mu_i = 0$ for all $i \geq 0$. We will show that for all $n \geq 1$ and $t \geq 0$ we have

$$P_t(1, n) = e^{-n\lambda t}(1 - e^{-\lambda t})^{n-1} \tag{1.4}$$

We do a proof by induction. Let $n = 1$. The only jump that may occur is to 2, the rate is λ. Hence, $P_t(1, 1)$ is the probability that the jump did not occur by time t. Therefore,

$$P_t(1, 1) = e^{-\lambda t},$$

and (1.4) holds for $n = 1$. Assume now that (1.4) holds for n. By (1.3) we have

$$P_t(1, n+1) = e^{-(n+1)\lambda t} \int_0^t e^{(n+1)\lambda s} n\lambda P_s(1, n) ds.$$

We now use the induction hypothesis for $P_s(1, n)$ to get

$$P_t(1, n+1) = e^{-(n+1)\lambda t} \int_0^t e^{(n+1)\lambda s} n\lambda e^{-\lambda s}(1 - e^{-\lambda s})^{n-1} ds.$$

Note that

$$e^{(n+1)\lambda s} e^{-\lambda s}(1 - e^{-\lambda s})^{n-1} = (e^{\lambda s} - 1)^{n-1} e^{\lambda s}.$$

Hence,

$$P_t(1, n+1) = e^{-(n+1)\lambda t} \int_0^t n\lambda(e^{\lambda s} - 1)^{n-1} e^{\lambda s} ds.$$

The integral is now easy to compute and

$$P_t(1, n+1) = e^{-(n+1)\lambda t}(e^{\lambda t} - 1)^n.$$

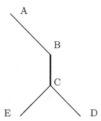

Fig. 9.1 This is an example of tree simulated using a Yule process. Species A gave birth to species B. Species B gave birth to species C. Species C gave birth to species D and E

Hence, (1.4) holds for $n + 1$ and therefore the formula is proved by induction.

This process was introduced by Yule in 1924 to model speciation (i.e., the birth of new species). Starting with one species $P_t(1, n)$ is the probability that the original species has given birth to n species by time t. This model also allows to simulate trees that keep track of which species gave birth to which, see Fig. 9.1. The parameter λ is interpreted as a mutation rate (a species gives birth to a new species at rate λ).

Yule model is surprisingly successful (given how simple it is) at explaining evolutionary patterns. It continues to be used to this day, see Nee (2006) for a review.

1.3 The Yule Process with Mutations

We use a modified Yule process to model the growth of pathogens in a human body before drug treatment starts. We modify the process in the following way. Every time there is a birth there is a fixed probability $\mu > 0$ that the new individual carry a given drug resistant mutation. We will compute the probability that the drug mutation has not occurred by time T. We think of T as the starting time for drug treatment.

Let A be the event that no mutation has appeared by time T. Assume that T is a fixed deterministic time. Assume that we start the process with a single individual. Conditioning on the number of individuals at time T we have

$$P(A|X(T) = n) = (1 - \mu)^{n-1}.$$

This is so because we have $n - 1$ individuals that were born by time T and each can acquire the mutation independently of the other individuals. Hence,

$$P(A|X(0) = 1) = \sum_{n \geq 1} P(A|X(T) = n) P(X(T) = n|X(0) = 1)$$

$$= \sum_{n \geq 1} (1 - \mu)^{n-1} e^{-\lambda T} (1 - e^{-\lambda T})^{n-1}.$$

Recall the geometric series for $|x| < 1$

$$\sum_{k=0}^{\infty} x^k = \frac{1}{1-x}.$$

We use the series for $x = (1 - \mu)(1 - e^{-\lambda T})$ to get

$$P(A|X(0) = 1) = \frac{e^{-\lambda T}}{1 - (1 - e^{-\lambda T})(1 - \mu)}.$$

It is easy to see that

$$P(A|X(0) = 1) \sim \frac{1}{\mu} e^{-\lambda T}$$

as T goes to infinity. That is, the probability of no drug resistant mutation decreases exponentially fast with T. This confirms the idea that treatment should start as early as possible.

1.4 Passage Times

Consider a continuous time birth and death chain. What is the expected time it takes for the chain to go from one state to another?

Assume that the birth and death rates are λ_i and μ_i. We assume that these rates are all strictly positive except for $\mu_0 = 0$. We are interested in computing the time it takes to go from i to j, $i < j$. Since a birth and death chain moves only one step at the time, it is enough to compute the time it takes to go from i to $i + 1$, from $i + 1$ to $i + 2, \ldots$ and from $j - 1$ to j and then sum these times. Let M_i be the expected time to go from state i to state $i + 1$. We have

$$M_0 = \frac{1}{\lambda_0}.$$

This is so because we assume that $\mu_0 = 0$. For $i \geq 1$, after an exponential time with rate $\lambda_i + \mu_i$ the chain jumps. Hence, the expected waiting time to jump is $\frac{1}{\lambda_i + \mu_i}$. It jumps to $i + 1$ with probability $\frac{\lambda_i}{\lambda_i + \mu_i}$ or to $i - 1$ with probability $\frac{\mu_i}{\lambda_i + \mu_i}$. If the chain jumps to the right, then it reaches $i + 1$. If it jumps to the left, it reaches $i - 1$. In order to get to $i + 1$ the chain has to go to i first. Thus, conditioning on the first jump we have

$$M_i = \frac{\lambda_i}{\mu_i + \lambda_i} \frac{1}{\mu_i + \lambda_i} + \frac{\mu_i}{\mu_i + \lambda_i} \left(\frac{1}{\mu_i + \lambda_i} + M_{i-1} + M_i \right).$$

Solving for M_i we get

$$M_i = \frac{1}{\lambda_i} + \frac{\mu_i}{\lambda_i} M_{i-1} \text{ for } i \geq 1 \tag{1.5}$$

We now apply this formula.

Example 1.5.   What is the expected time to go from 0 to 2?
   We have

$$M_0 = \frac{1}{\lambda_0}.$$

We use (1.5) to get

$$M_1 = \frac{1}{\lambda_1} + \frac{\mu_1}{\lambda_1} M_0.$$

The expected time to go from 0 to 2 is

$$M_0 + M_1 = \frac{1}{\lambda_0} + \frac{1}{\lambda_1} + \frac{\mu_1}{\lambda_1} \frac{1}{\lambda_0}.$$

Example 1.6.   Assume that $\lambda_i = \lambda$ for all $i \geq 0$ and $\mu_i = \mu$ for all $i \geq 1$.
Then (1.5) becomes

$$M_i = \frac{1}{\lambda} + \frac{\mu}{\lambda} M_{i-1} \text{ for } i \geq 1.$$

Since

$$M_{i-1} = \frac{1}{\lambda} + \frac{\mu}{\lambda} M_{i-2} \text{ for } i \geq 2$$

we have

$$M_i - M_{i-1} = \frac{\mu}{\lambda} M_{i-1} - \frac{\mu}{\lambda} M_{i-2}$$

for $i \geq 2$. Let $a_i = M_i - M_{i-1}$, then

$$a_i = \frac{\mu}{\lambda} a_{i-1} \text{ for all } i \geq 2.$$

Thus,

$$a_i = (\frac{\mu}{\lambda})^{i-1} a_1 \text{ for all } i \geq 1.$$

Note that $a_1 = M_1 - M_0 = \frac{\mu^2}{\lambda}$. Hence,

$$M_i - M_{i-1} = \frac{\mu^i}{\lambda^{i+1}} \text{ for all } i \geq 1.$$

Note that

$$\sum_{i=1}^{j}(M_i - M_{i-1}) = M_j - M_0 = \sum_{i=1}^{j} \frac{\mu^i}{\lambda^{i+1}}.$$

Assuming that $\mu \neq \lambda$ and summing the geometric sequence we get

$$M_j = \frac{1}{\lambda} + \frac{\mu}{\lambda^2} \frac{1 - (\frac{\mu}{\lambda})^j}{1 - \frac{\mu}{\lambda}} \text{ for all } j \geq 0.$$

Problems

1. Consider a birth and death chain on $\{0, 1\}$. The only λ_i and μ_i that are non-zero are λ_0 and μ_1. We simplify the notation by letting $\lambda = \lambda_0$ and $\mu = \mu_1$.

(a) Write the backward Kolmogorov equations for this process.
(b) From (a), get that

$$(P_t(0,0) - P_t(1,0))' = -(\lambda + \mu)(P_t(0,0) - P_t(1,0)).$$

(c) Using (b), compute $P_t(0,0) - P_t(1,0)$ and then $P_t(0,0)$, $P_t(1,0)$, $P_t(1,1)$, and $P_t(0,1)$.
(d) Compute $\lim_{t\to\infty} P_t(i,j)$ for $i, j = 0, 1$.

2. Consider a queue with a single server. Assume that the interarrival times of customers are exponentially distributed with rate 2 and assume that the service time is exponentially distributed with rate μ. Show that this is a birth and death chain and find the birth and death rates λ_i and μ_i.

3. Same question as in Problem 2 with infinitely many servers and rate μ per service time per server.

4. Write the forward and backward equations for Example 1.1.

5. Write the forward and backward equations for a queue with infinitely many servers, arrival rate λ and service rate μ.

6. Consider a Yule process with $\lambda = 1$. Compute $P_t(0,0)$, $P_t(1,0)$, $P_t(0,1)$, and $P_t(1,2)$.

7. Consider a birth and death chain. Compute the backward differential equation for the transition probability $P_t(0, j)$ where $j \geq 0$.

8. Consider the Poisson process with birth rate λ. That is, $\lambda_i = \lambda$ and $\mu_i = 0$ for all $i \geq 0$. Use the method of Sect. 1.1. to compute the transition probabilities $P_t(0, i)$ for all $i \geq 0$.

9. Write the backward equations for the Poisson process.

10. Consider a birth and death chain with $\lambda_n = n\lambda$ and $\mu_n = n\mu$. Let $M(t) = E(X(t))$.

(a) Show that

$$\frac{d}{dt} M(t) = (\lambda - \mu) M(t).$$

(b) Find $M(t)$.

11. Consider the Yule process. Show that for $i \geq 1$ and $j \geq i$ we have

$$P_t(i, j) = \binom{j-1}{j-i} e^{-i\lambda t} (1 - e^{-\lambda t})^{j-i}.$$

Show that this formula is correct by checking that it satisfies the differential equations

$$\frac{d}{dt} P_t(i, j) = \lambda_{j-1} P_t(i, j-1) - \lambda_j P_t(i, j).$$

12. In Example 1.6, show that if $\lambda = \mu$ then

$$M_j = \frac{j+1}{\mu} \quad \text{for all } j \geq 0.$$

13. In Example 1.6 compute the time it takes to go from 1 to 5 in the following cases

(a) $\lambda = \mu = 1$.
(b) $\lambda = 2, \mu = 1$.
(c) $\lambda = 1, \mu = 2$.

14. Consider a birth and death chain. Compute the expected time it takes to go from 1 to 0 as a function of the λ_i's and μ_i's.

15. Use a Yule process with $\lambda = 1$ to simulate an evolutionary tree such as the one in Fig. 9.1.

16. Prove Lemma 1.1.

2   Limiting Probabilities

Recall that π is a probability distribution on the positive integers if

$$\pi(i) \geq 0 \text{ for all } i \geq 0 \text{ and } \sum_{i=0}^{\infty} \pi(i) = 1.$$

Definition 2.1. Let π be a probability distribution on the positive integers. Then π is a stationary distribution for a continuous time chain with transition probabilities $P_t(i, j)$ if

$$\sum_{j \geq 0} \pi(j) P_t(j, k) = \pi(k) \text{ for all states } i, j, \text{ and all times } t \geq 0.$$

Example 2.1. Consider a chain $(X_t)_{t \geq 0}$ with transition probabilities P_t and stationary distribution π. Assume that the chain starts in a state picked according to the stationary distribution π. What is the probability that at time t the chain is in state k?

By conditioning on the initial state we have

$$P_\pi(X_t = k) = \sum_{i \geq 0} P(X_t = k | X_0 = i) P(X_0 = i).$$

By assumption $P(X_0 = i) = \pi(i)$. Hence, using that π is stationary

$$P_\pi(X_t = k) = \sum_{i \geq 0} \pi(i) P_t(i, k) = \pi(k).$$

Therefore, the distribution of X_t is π for every $t \geq 0$. It does not depend on t. This is why it is called a stationary distribution.

Definition 2.2. A chain is said to be irreducible if its transition probabilities $P_t(i, j)$ are strictly positive for all $t > 0$ and all states i and j.

In words an irreducible chain can go from any state to another in any fixed time. This is not a strong constraint. It can be shown that either $P_t(i, j) > 0$ for all $t > 0$ or $P_t(i, j) = 0$ for all $t > 0$. In particular, for a birth and death chain with $\mu_0 = 0$, $\mu_i > 0$ for all $i \geq 1$ and $\lambda_i > 0$ for all $i \geq 0$ this condition holds. See the problems.

We now state the main theorem of this section. The proof can be found in more advanced books such as Feller (1971) (see Sect. XIV.9).

Theorem 2.1. *Consider an irreducible continuous time chain on the positive integers. For all $i \geq 0$ and $j \geq 0$ the following limit exists.*

$$\lim_{t \to \infty} P_t(i, j) = L(j).$$

Moreover, there are two possibilities, either

(i) $L(j) = 0$ for all j
   or
(ii) $L(j) > 0$ for all j and L is then the unique stationary distribution of the chain.

In words, Theorem 2.1 tells us that for an irreducible chain $P_t(i, j)$ always converges. The limits are all zero or all strictly positive. Moreover, in the latter case the limits form the (unique) stationary distribution. Note also that the limit of $P_t(i, j)$ does not depend on the initial state i.

Example 2.2. Consider the Poisson process $N(t) \geq 0$ with rate $\lambda > 0$. Note that this is not an irreducible chain (why?). However, the limits of $P_t(i, j)$ exist and can be computed. We concentrate on $P_t(0, n)$ since the Poisson process starts at 0 by definition. We know that

$$P_t(0, n) = P(N(t) = n) = e^{-\lambda t} \frac{(\lambda t)^n}{n!}.$$

Recall that for all x we have power series expansion

$$e^x = \sum_{n=0}^{\infty} \frac{x^n}{n!}.$$

Hence, for any $n \geq 0$ and all $x \geq 0$ we have

$$e^x \geq \frac{x^{n+1}}{(n+1)!}.$$

Therefore,

$$\frac{e^x}{x^n} \geq \frac{x}{(n+1)!}$$

and

$$x^n e^{-x} \leq \frac{(n+1)!}{x}.$$

Thus, for any fixed $n \geq 0$ as x goes to infinity we have

$$\lim_{x \to +\infty} x^n e^{-x} = 0.$$

We apply this limit to $x = \lambda t$ to get

$$\lim_{t \to +\infty} P_t(0, n) = 0.$$

This limit is not really surprising given that this is a pure birth process with constant birth rate. We expect the Poisson process to go to infinity as t goes to infinity.

As the next result shows finite chains always have a stationary distribution.

Corollary 2.1. *Consider an irreducible continuous time chain on a finite set. Then, the chain has a unique stationary distribution π. Moreover,*

$$\lim_{t \to \infty} P_t(i, j) = \pi(j) \text{ for all } i, j.$$

Proof of Corollary 2.1. According to Theorem 2.1 there is a limit for each $P_t(i, j)$ that we denote by $\pi(j)$. We also have

$$\sum_j P_t(i, j) = 1.$$

This is a finite chain so the preceding sum is finite and we may interchange sum and limit to get

$$\lim_{t \to \infty} \sum_j P_t(i, j) = \sum_j \lim_{t \to \infty} P_t(i, j) = \sum_j \pi(j) = 1.$$

This is the critical point of the proof and it depends heavily on having a finite chain. Interchanging infinite sums and limits is not always allowed. Since $\sum_j \pi(j) = 1$ at least one of the $\pi(j)$ is strictly positive. Thus, by Theorem 2.1 all the $\pi(j)$ are strictly positive. Therefore, π is a stationary distribution for the chain.

By Theorem 2.1 we know that the stationary distribution (if it exists) is unique. In the case of finite chains the proof of uniqueness is simple so we give it now. Assume there is another stationary distribution $a(j)$. Then by definition

$$\sum_j a(j) P_t(j, k) = a(k).$$

Letting t go to infinity on both sides yields

$$\lim_{t \to +\infty} \sum_j a(j) P_t(j, k) = a(k).$$

Again since this is a finite sum we may interchange limit and sum to get

$$\sum_j a(j) \pi(k) = a(k).$$

Since

$$\sum_j a(j)\pi(k) = \pi(k)\sum_j a(j)$$

and since the sum of the $a(j)$ is 1 we get that $\pi(k) = a(k)$ for every k. This proves that there is a unique stationary distribution and concludes the proof of Corollary 2.1.

We now turn to the question of computing the stationary distribution for birth and death chains. In most examples $P_t(i, j)$ is not known explicitly so we cannot use Theorem 2.1 to find π. Instead we use the Kolmogorov differential equations. Recall that the Kolmogorov forward differential equations are

$$P_t'(i, j) = \lambda_{j-1}P_t(i, j-1) - (\lambda_j + \mu_j)P_t(i, j) + \mu_{j+1}P_t(i, j+1) \text{ for } j \geq 1 \text{ and } i \geq 0$$

and

$$P_t'(i, 0) = -\lambda_0 P_t(i, 0) + \mu_1 P_t(i, 1) \text{ for } i \geq 0.$$

Letting t go to infinity in the forward equations and assuming that

$$\lim_{t\to\infty} P_t(i, j) = \pi(j)$$

$$\lim_{t\to\infty} P_t'(i, j) = 0,$$

we get the following.

Theorem 2.2. *A birth and death chain with rates λ_i and μ_i has a stationary distribution π if and only if π is a solution of*

$$0 = \lambda_{j-1}\pi(j-1) - (\lambda_j + \mu_j)\pi(j) + \mu_{j+1}\pi(j+1) \text{ for } j \geq 1$$

$$0 = -\lambda_0\pi(0) + \mu_1\pi(1)$$

and

$$\sum_j \pi(j) = 1.$$

A good way to remember these equations is to write that when the process is in steady state the rate at which the process enters state j is equal to the rate at which the process leaves state j:

$$\lambda_{j-1}\pi(j-1) + \mu_{j+1}\pi(j+1) = (\lambda_j + \mu_j)\pi(j) \tag{2.1}$$

See Liggett (2010) for a proof of Theorem 2.2. We will now solve this system of equations to find π and the condition of existence for π. From

$$0 = \lambda_{j-1}\pi(j-1) - (\lambda_j + \mu_j)\pi(j) + \mu_{j+1}\pi(j+1) \text{ for } j \geq 1$$

we get for $j \geq 1$

$$\mu_{j+1}\pi(j+1) - \lambda_j\pi(j) = \mu_j\pi(j) - \lambda_{j-1}\pi(j-1) \qquad (2.2)$$

Let

$$a_j = \mu_j\pi(j) - \lambda_{j-1}\pi(j-1).$$

By (2.2) we see that for all $j \geq 1$ we have $a_{j+1} = a_j$. Hence, $a_j = a_{j-1}, a_{j-1} = a_{j-2}$ and so on. Therefore,

$$a_{j+1} = a_1.$$

That is, for $j \geq 1$

$$\mu_{j+1}\pi(j+1) - \lambda_j\pi(j) = \mu_1\pi(1) - \lambda_0\pi(0).$$

Note now that $a_1 = \mu_1\pi(1) - \lambda_0\pi(0)$. Since π is stationary $a_1 = 0$. Therefore, $a_j = 0$ for all $j \geq 1$. Hence,

$$\pi(j) = \frac{\lambda_{j-1}}{\mu_j}\pi(j-1).$$

We now show by induction that for all $j \geq 1$

$$\pi(j) = \frac{\lambda_0 \dots \lambda_{j-1}}{\mu_1 \dots \mu_j}\pi(0).$$

Since $a_1 = 0$ the formula holds for $j = 1$. Assume now that the formula holds for j. Since $a_{j+1} = 0$ we get

$$\pi(j+1) = \frac{\lambda_j}{\mu_{j+1}}\pi(j).$$

Using the induction hypothesis

$$\pi(j+1) = \frac{\lambda_j}{\mu_{j+1}}\frac{\lambda_0 \dots \lambda_{j-1}}{\mu_1 \dots \mu_j}\pi(0).$$

This completes the induction proof.

We also need

$$\sum_{j \geq 0} \pi(j) = 1.$$

That is,

$$\pi(0)\Big(1 + \sum_{j \geq 1} \frac{\lambda_0 \ldots \lambda_{j-1}}{\mu_1 \ldots \mu_j}\Big) = 1.$$

This equation has a solution $\pi(0) > 0$ if and only if

$$\sum_{j \geq 1} \frac{\lambda_0 \ldots \lambda_{j-1}}{\mu_1 \ldots \mu_j} < \infty.$$

We have proved the following.

Proposition 2.1. *Consider a birth and death chain with rates $\mu_0 = 0$, $\mu_i > 0$ for all $i \geq 1$ and $\lambda_i > 0$ for all $i \geq 0$. The chain has a stationary distribution if and only if*

$$C = \sum_{j \geq 1} \frac{\lambda_0 \ldots \lambda_{j-1}}{\mu_1 \ldots \mu_j} \text{ is finite.}$$

Then the stationary distribution is given by

$$\pi(0) = \frac{1}{C + 1}$$

$$\pi(j) = \frac{\lambda_0 \ldots \lambda_{j-1}}{\mu_1 \ldots \mu_j} \pi(0) \text{ for } j \geq 1.$$

Proposition 2.1 applies to birth and death chains on infinite sets. We already knew by Corollary 2.1 that irreducible birth and death chains on finite sets always have a unique stationary distribution. We now turn to a few examples.

Example 2.3. Consider the infinite server queue. Assume that the interarrival times have rate λ. Assume that there are infinitely many servers and that the rate of service per server is μ. Let X_t be the number of customers in the queue.

If $X_t = n$, then the chain jumps to $n + 1$ with rate λ or it jumps to $n - 1$ (if $n \geq 1$) with rate $n\mu$. This is so because in order to jump to $n - 1$ one of the n servers must finish their job and we know that the minimum of n independent exponential distributions with rate μ is an exponential distribution with rate $n\mu$. Thus, this is an irreducible birth and death chain on the positive integers with rates:

$$\lambda_i = \lambda \text{ and } \mu_i = i\mu \text{ for } i \geq 0.$$

We have

$$\frac{\lambda_0 \ldots \lambda_{j-1}}{\mu_1 \ldots \mu_j} = \frac{\lambda^j}{\mu^j j!}.$$

Recall that for all x

$$\sum_{k \geq 0} \frac{x^k}{k!} = e^x,$$

so

$$C + 1 = 1 + \sum_{j \geq 1} \frac{\lambda^j}{\mu^j j!} = e^{\lambda/\mu} < \infty.$$

Hence, by Proposition 2.1 the infinite server queue has a stationary distribution π for all strictly positive values of λ and μ. And π is given by

$$\pi(j) = \frac{\lambda^j}{\mu^j j!} e^{-\lambda/\mu}.$$

That is, π is a Poisson distribution with parameter λ/μ.

Example 2.4. Our second example is a birth and death chain on a finite set.

Consider a job shop with two machines and one serviceman. Suppose that the amount of time each machine runs before breaking down is exponentially distributed with rate λ. Suppose the time it takes for a serviceman to fix a machine is exponentially distributed with rate μ. Assume that these exponential times are independent.

We say that the system is in state n when n machines are broken. This is a birth and death chain on $\{0, 1, 2\}$. For $0 \leq n \leq 2$, μ_n is the rate at which we go from n to $n - 1$ broken machines and λ_n is the rate at which we go from n to $n + 1$ broken machines. Since there is only one serviceman we have

$$\mu_n = \mu \text{ for } 1 \leq n \leq 2.$$

Since there are two machines

$$\lambda_n = (2 - n)\lambda \text{ for } 0 \leq n \leq 2.$$

This is an irreducible finite chain. Therefore it has a stationary distribution. By (2.1) we have

$$\mu\pi(1) = 2\lambda\pi(0)$$
$$\mu\pi(2) + 2\lambda\pi(0) = (\lambda + \mu)\pi(1)$$
$$\lambda\pi(1) = \mu\pi(2).$$

Solving this system yields

$$\pi(0) = \frac{\mu}{2\lambda} \frac{1}{1 + \lambda/\mu + \mu/2\lambda}$$

$$\pi(1) = \frac{1}{1 + \lambda/\mu + \mu/2\lambda}$$

$$\pi(2) = \frac{\lambda}{\mu} \frac{1}{1 + \lambda/\mu + \mu/2\lambda}.$$

Problems

1. Consider a single server queue with arrival rate λ and service rate μ.

(a) Show that the stationary distribution exists if and only if $\lambda/\mu < 1$. Find the stationary distribution under this condition.

(b) What is the average length of the queue after a long time?

2. Consider an N server queue with arrival rate λ and service rate μ.

(a) Find a condition on λ, μ and N which ensures the existence of a stationary distribution.

(b) What is the probability that there are exactly two individuals in the queue after a long time, under the condition found in (a)?

3. Consider a queue with two servers and arrival rate λ. Assume that a maximum of three people can be in the system. Assume that the service rate per server is μ.

(a) Explain why you know without doing any computation that there is a stationary distribution.

(b) Find the stationary distribution.

4. Show that a pure birth process has no stationary distribution. (Show that the only solution of the equations in Theorem 2.2 is π identically 0.)

5. Consider a walker on the positive integers. After a mean 1 exponential time the walker jumps one unit to the left with probability q or one unit to the right with probability p where $p + q = 1$. When the walker is at 0 it jumps to 1 (after a mean 1 exponential time). See Example 1.1.

(a) Show that this process has a stationary distribution if and only if $p < 1/2$.

(b) What happens to the walker when $p > 1/2$?

6. Consider a population where each individual gives birth with rate λ or dies with rate μ. Moreover, there is a constant rate θ of immigration if the population is below N. If the population is above N, no immigration is allowed.

(a) Find a condition on λ, μ, N, and θ in order to have a stationary distribution for the size of the population.
(b) Assume that $N = 3$, $\theta = \lambda = 1$, and $\mu = 2$. What is the proportion of time that immigration is not allowed?

7. A car wash has room for at most two cars (including the one being washed). So if there are already two cars in the system the following potential customers do not wait and leave. Potential customers arrive at rate 3 per hour. The service time rate is 4 per hour.

(a) In the steady state, what is the expected number of cars in the car wash?
(b) What is the proportion of potential customers that enter the car wash?
(c) If the car wash were twice as fast, how much more business would it do?

8. Consider a job shop with two machines and two servicemen. Suppose that the amount of time each machine runs before breaking down is exponentially distributed with rate 10. Suppose the time it takes for a serviceman to fix a machine is exponentially distributed with rate 3. Assume that all these exponential times are independent.

(a) What is the mean number of machines in use in the long run?
(b) What proportion of time are the two servicemen busy in the long run?

9. Consider a taxi station where interarrival times for taxis and customers, are respectively, 1 and 3 per minute. Assume that taxis wait in line for customers but that customers do not wait for taxis.

(a) What is the mean number of taxis waiting in the long run?
(b) What is the probability for a customer to find a taxi in the long run?

10. Consider a single server queue with arrival rate λ and service rate μ. Assume that $\lambda < \mu$.

(a) Show that this chain has a stationary distribution

$$\pi(n) = (\frac{\lambda}{\mu})^n(1 - \frac{\lambda}{\mu}) \text{ for } n \geq 0.$$

(b) Consider the chain in its stationary state. Let T be the waiting time of an arriving customer including his own service time. If there are n customers at the arriving time, show that T is the sum of $n+1$ independent exponential random variables.
(c) Recall that a sum of $n + 1$ independent exponential distributions with rate μ is a Γ distribution with parameters $n + 1$ and μ. The density of $\Gamma(n + 1, \mu)$ is given by

$$\frac{1}{n!}\mu^{n+1}t^n e^{-\mu t} \text{ for } t \geq 0.$$

Compute $P(T \leq t)$ by conditioning on the number of customers in the queue at the customer arrival time.

11. Let $t > 0$ and $i \geq 1$.

(a) Show that

$$P_t(i, i + 1) \geq (1 - e^{-\lambda_i t}) e^{-\mu_i t} e^{-\mu_{i+1} t} e^{-\lambda_{i+1} t}.$$

(One way to be in $i + 1$ at time t is to have a jump forward from i to $i + 1$, no jump backward from i to $i - 1$ and to stay put at $i + 1$.)

(b) Show that $P_t(i, i + 1) > 0$.

(c) Show that $P_t(i, i - 1) > 0$.

(d) Show that a birth and death chain with $\mu_0 = 0$, $\mu_i > 0$ for all $i \geq 1$ and $\lambda_i > 0$ for all $i \geq 0$ is irreducible.

Notes

In order for a birth and death chain to be properly defined the rates λ_n cannot increase too rapidly with n. The question of existence and construction of continuous time Markov chains involves higher mathematics (in particular measure theory). Good books on the subject are Bhattacharya and Waymire (1990), Feller (1971), or Liggett (2010). Here we concentrate on Markov chains on finite or countable sets. Things get more complicated for chains on uncountable sets. See, for instance, the voter model in Liggett (2010).

References

Bhattacharya, R.N., Waymire, E.C.: Stochastic Processes with Applications. Wiley, New York (1990)

Feller, W.: An Introduction to Probability Theory and Its Applications, vol. 2, 2nd edn. Wiley, New York (1971)

Liggett, T.: Continuous Time Markov Processes. Graduate Studies in Mathematics, vol. 113. American Mathematical Society, Providence (2010)

Nee, S.: Birth-death models in macroevolution. Annu. Ecol. Evol. Syst. **37**, 1–17 (2006)

Chapter 10
Percolation

Percolation is the first spatial model we will consider. The model is very easy to define but not so easy to analyze. Elementary methods can however be used to prove a number of results. We will use combinatorics, discrete branching processes, and coupling techniques.

1 Percolation on the Lattice

We start with some definitions. Let $\mathbf{Z}^d$ be the d dimensional lattice. We say that $x = (x_1, x_2, \ldots, x_d)$ and $y = (y_1, y_2, \ldots, y_d)$ in $\mathbf{Z}^d$ are nearest neighbors if

$$\sum_{i=1}^{d} |x_i - y_i| = 1.$$

Let p be a real number in $[0, 1]$. For each $x \in \mathbf{Z}^d$ there are $2d$ **edges** linking x to each of its $2d$ nearest neighbors. We declare each edge open with probability p and closed otherwise, independently of all other edges (see Fig. 10.1.)

We say that there is a **path** between x and y if there is a sequence $(x_i)_{0 \le i \le n}$ of sites in $\mathbf{Z}^d$ such that $x_0 = x$, $x_n = y$, x_i and x_{i+1} are nearest neighbors for $0 \le i \le n-1$. The path is said to be **open** if all the edges x_i, x_{i+1} are open for $0 \le i \le n-1$.

For a fixed x in $\mathbf{Z}^d$ let $C(x)$ be the (random) set of y's such that there is an open path from x to y. We call $C(x)$ the **open cluster** of x. By convention we assume that x belongs to $C(x)$. Note that the distribution of $C(x)$ is the same for all x. Hence, we concentrate on $C(0)$ that we also denote by C. Let $|C|$ be the number of elements in C.

© Springer Science+Business Media New York 2014
R.B. Schinazi, *Classical and Spatial Stochastic Processes: With Applications to Biology*, DOI 10.1007/978-1-4939-1869-0_10

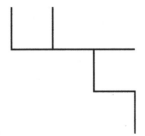

Fig. 10.1 This is part of the percolation model on $\mathbf{Z}^2$. *Solid lines* indicate open edges

Definition 1.1. Let $\theta(p)$ be the probability that the origin is in an infinite cluster. That is, $\theta(p) = P(|C| = \infty)$. If $\theta(p) > 0$, we say that there is percolation.

Observe that if $p = 0$ then all edges are closed and therefore $\theta(0) = 0$. On the other hand, if $p = 1$, then all edges are open and $\theta(1) = 1$.

Proposition 1.1. *The function θ is increasing.*

Proof of Proposition 1.1. We use a coupling argument. We construct simultaneously percolation models for all $p \in [0, 1]$. For each edge e in $\mathbf{Z}^d$ let $U(e)$ be a uniform random variable on $[0,1]$. Moreover, we take all the $U(e)$ to be mutually independent. We declare an edge e to be open for the percolation model with parameter p if and only if $U(e) < p$. Since $P(U(e) < p) = p$ an edge is open with probability p and closed with probability $1 - p$.

This gives a simultaneous construction for all percolation models (i.e., for all p in $[0, 1]$). In particular, take $p_1 < p_2$ and note that if $U(e) < p_1$ then $U(e) < p_2$. Thus, letting C_p denote the open cluster of the origin for the model with parameter p, we get

$$C_{p_1} \subset C_{p_2} \text{ for } p_1 < p_2.$$

So

$$\{|C_{p_1}| = \infty\} \subset \{|C_{p_2}| = \infty\} \text{ for } p_1 < p_2.$$

Hence,

$$\theta(p_1) = P(|C_{p_1}| = \infty) \le \theta(p_2) = P(|C_{p_2}| = \infty) \text{ for } p_1 < p_2.$$

This shows that θ is an increasing function of p and completes the proof of Proposition 1.1.

Let $p_c(d)$ be the **critical value** of the percolation model on $\mathbf{Z}^d$. We define it by

$$p_c(d) = \sup\{p : \theta(p) = 0\}.$$

Since θ is an increasing function, if $p < p_c(d)$ then $\theta(p) = 0$ (there is no percolation) while if $p > p_c(d)$ then $\theta(p) > 0$ (there is percolation). In other words, if $p < p_c(d)$, the origin is in a finite cluster with probability one. If $p > p_c(d)$, the origin has a strictly positive probability of being in an infinite cluster. Both behaviors actually occur if $p_c(d)$ is not 0 or 1. If $p_c(d)$ is in $(0,1)$, the model is said to exhibit a nontrivial phase transition. What makes percolation an interesting model is that there is a phase transition for all $d \geq 2$ (but not for $d = 1$). We will now show this for $d = 2$, see the problems for $d \neq 2$.

Theorem 1.1. *We have a nontrivial phase transition for percolation on* $\mathbf{Z}^2$ *in the sense that* $0 < p_c(2) < 1$. *More precisely, we have the following bounds*

$$1/3 \leq p_c(2) \leq \frac{11 + \sqrt{13}}{18}.$$

It turns out that $p_c(2)$ is exactly 1/2 (see Grimmett 1999). The proof is very involved so we will only show the rather crude bounds stated in Theorem 1.1. No other $p_c(d)$ is known exactly and even good bounds are very hard to get.

Proof of Theorem 1.1. We start by showing that $p_c(2) \geq 1/3$. It is enough to show that for all $p < 1/3$ we have $\theta(p) = 0$ (why?).

Recall that a path is defined as being a sequence of nearest neighbor sites in $\mathbf{Z}^d$. A **self-avoiding** path is a path for which all sites in the path are distinct. Let $S(n)$ be the total number of self-avoiding paths starting at the origin and having length n. The exact value of $S(n)$ becomes very difficult to compute as n increases. We will compute an upper bound instead. Observe that starting at the origin of $\mathbf{Z}^2$ there are four choices for the second site of the self-avoiding path. For the third site on of the path there is at most three possible choices since the path is self-avoiding. Hence, $S(n)$ is less than $4(3)^{n-1}$.

Let $N(n)$ be the (random) number of open self-avoiding paths starting at the origin and having length n. The random variable $N(n)$ can be represented as

$$N(n) = \sum_{i=1}^{S(n)} X_i$$

where $X_i = 1$ if the i-th self-avoiding path is open and $X_i = 0$ otherwise. Since a path of length n has probability p^n of being open we have

$$P(X_i = 1) = p^n.$$

Hence,

$$E(N(n)) = S(n)p^n \leq 4(3)^{n-1}p^n.$$

Note that if the origin is in an infinite open cluster there must exist open paths of all lengths starting at the origin. Thus for all n

$$\theta(p) \leq P(N(n) \geq 1).$$

Note that

$$E(N(n)) = \sum_{k \geq 1} k P(N(n) = k) \geq \sum_{k \geq 1} P(N(n) = k) = P(N(n) \geq 1).$$

Thus, for every $n \geq 1$ we have

$$\theta(p) \leq P(N(n) \geq 1) \leq E(N(n)) \leq p^n 4(3)^{n-1}.$$

Take $p < \frac{1}{3}$ and let n go to infinity in the preceding inequality to get $\theta(p) = 0$. This proves that

$$p_c(2) \geq \frac{1}{3}.$$

This gives the lower bound in Theorem 1.1.

We now deal with the upper bound. It turns out that $\mathbf{Z}^2$ has some special properties that make the analysis much easier than for $\mathbf{Z}^d$ when $d > 2$. We define a dual graph of $\mathbf{Z}^2$ by

$$\{x + (1/2, 1/2) : x \in \mathbf{Z}^2\}.$$

We join two such nearest neighbor vertices by a straight line. The dual graph looks exactly like $\mathbf{Z}^2$ (and this is particular to dimension 2). We declare an edge in the dual to be open if it crosses an open edge in $\mathbf{Z}^d$. We declare an edge in the dual to be closed if it crosses a closed edge in $\mathbf{Z}^d$. A **circuit** is a path that ends at its starting point. A circuit is self-avoiding except that the first and last points are the same. The crucial remark in this proof is the following. The open cluster of the origin is finite if and only if there is a closed circuit in the dual that surrounds the origin (see Fig. 10.2).

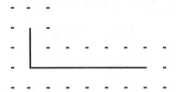

Fig. 10.2 A finite open cluster (*solid lines*) surrounded by a closed circuit (*dotted lines*) in the dual lattice

Define

$A_n = \{$ there is a closed circuit of length n in the dual surrounding the origin$\}$.

Then

$$\{|C| < \infty\} = \bigcup_{n \geq 1} A_n$$

and so

$$P(|C| < \infty) \leq \sum_{n \geq 1} P(A_n).$$

We now count the closed circuits of length n. If a circuit surrounds the origin, then it must pass through some vertex $(k + 1/2, 1/2)$ where $k \geq 0$. Since the circuit has length n we must have $k < n$. Otherwise the circuit would be longer than $2n$. Hence, there is at most n choices for the starting vertex $(k + 1/2, 1/2)$. We have three choices for each of the $n - 1$ remaining vertices of the circuit. Therefore the number of circuits of length n surrounding the origin is at most $n3^{n-1}$. Moreover the probability that all edges of the circuit are closed is $(1 - p)^n$. Thus,

$$P(|C| < \infty) \leq \sum_{n \geq 1} n3^{n-1}(1 - p)^n.$$

Recall that

$$\sum_{n \geq 1} nx^{n-1} = \frac{1}{(1 - x)^2}.$$

Let

$$f(p) = \sum_{n \geq 1} n3^{n-1}(1 - p)^n = \frac{1 - p}{(-2 + 3p)^2}.$$

Observe that f is defined for p in $(2/3, 1]$ and that $f(p) < 1$ if and only if $p > \frac{11 + \sqrt{13}}{18}$. This proves that

$$P(|C| < \infty) \leq f(p) < 1 \text{ for } p > \frac{11 + \sqrt{13}}{18}.$$

So $p_c(2) \leq \frac{11 + \sqrt{13}}{18} < 1$ and this completes the proof of Theorem 1.1.

If $p > p_c$ we know that with positive probability the origin is in an infinite open cluster. But unless $p = 1$ we know that $\theta(p) < 1$ (why?), so there is also a positive probability that the origin is not in an infinite open cluster. Do we have an infinite open cluster somewhere else? The answer is yes.

Proposition 1.2. *If $p > p_c(d)$, the probability that there is an infinite open cluster somewhere in $\mathbf{Z}^d$ is one. If $p < p_c(d)$, the probability that there is an infinite open cluster somewhere in $\mathbf{Z}^2$ is zero.*

Proof of Proposition 1.2. Let A be the event "there is an open cluster somewhere in $\mathbf{Z}^d$." In order for A to happen at least one cluster $C(x)$ must be infinite. Thus,

$$P(A) \leq \sum_{x \in \mathbf{Z}^d} P(|C(x)| = \infty).$$

If $p < p_c(d)$, then by translation invariance of the model

$$P(|C(x)| = \infty) = P(|C| = \infty) = 0$$

for every x. Therefore, $P(A) = 0$ for $p < p_c(d)$.

On the other hand, for A to happen it is enough to have the origin in an infinite cluster. Thus,

$$P(A) \geq P(|C| = \infty) = \theta(p).$$

So, if $p > p_c(d)$ then $P(A) > 0$. The event A does not depend on any finite set of edges. That is, we may close or open a finite number of edges without changing the outcome: A occurs or A does not occur. Such an event is called a tail event. Since the states of the edges are mutually independent we may use Kolmogorov's 0–1 law (see, for instance, Durrett 2010) to deduce $P(A) = 0$ or 1. Thus, when $p > p_c(d)$, $P(A) = 1$. The proof of Proposition 1.2 is complete.

Problems

1. Consider the percolation model on $\mathbf{Z}$. Let C be the open cluster of the origin.

(a) Show that for all $n \geq 2$

$$P(|C| = n) \leq (n + 1)p^n.$$

(b) Show that

$$P(|C| \geq n) \leq \sum_{k \geq n}(k + 1)p^k.$$

(c) Show for $n \geq 1$ that

$$\theta(p) = P(|C| = \infty) \le P(|C| \ge n).$$

(d) Show that if $p < 1$ then $\theta(p) = 0$.

(e) Show that the critical value $p_c(1)$ is 1.

2. Consider the percolation model on $\mathbf{Z}^d$.

(a) Show that if $\theta(p_1) = 0$ for some p_1 then $p_c(d) \ge p_1$.

(b) Show that if $\theta(p_2) > 0$ for some p_2 then $p_c(d) \le p_2$.

3. Consider percolation on $\mathbf{Z}^d$ for $d \ge 2$.

(a) Show that

$$\theta(p) \le p^n (2d)(2d - 1)^{n-1}.$$

(Find an upper bound for the number of self-avoiding random walks such as in the proof of Theorem 1.1.)

(b) Show that

$$p_c(d) \ge \frac{1}{2d - 1}.$$

4. (a) Use uniform random variables $U(e)$ such as in the proof of Proposition 1.1 to construct the model on $\mathbf{Z}^{d+1}$.

(b) Show that the construction in (a) also provides a construction of percolation on $\mathbf{Z}^d$.

(c) Let $\theta_d(p)$ be the probability that the origin of $\mathbf{Z}^d$ is in an infinite open cluster. Show that for all p in $[0, 1]$ we have

$$\theta_d(p) \le \theta_{d+1}(p).$$

(d) Show that

$$p_c(d) \ge p_c(d + 1).$$

(e) Using (d) and Theorem 1.1 show that $p_c(d) < 1$ for all $d \ge 2$.

5. Show that the critical value $p_c(d)$ is in $(0, 1)$ for all dimensions $d \ge 2$. (Use Problems 3 and 4.)

6. Show that the percolation probability $\theta(p) = 1$ if and only if $p = 1$.

7. Let $S(n)$ be the total number of self-avoiding paths starting at the origin and having length n in $\mathbf{Z}^d$.

(a) Show that

$$S(n) \le 2d(2d - 1)^{n-1}.$$

(b) Show that

$$S(n) \geq d^n.$$

(A path that moves from site to site increasing by 1 one of its d coordinates is self-avoiding.)

(c) Use (a) and (b) to argue that $S(n)$ increases exponentially with n.

8. Let $S(n)$ be the total number of self-avoiding paths starting at the origin and having length n in $\mathbf{Z}^d$.

(a) Show that for n and m we have

$$S(n + m) \leq S(n)S(m).$$

(Observe that breaking a self-avoiding path in two pieces yields two self-avoiding paths.)

(b) Define $a(n) = \ln S(n)$. Show that

$$a(m + n) \leq a(n) + a(m).$$

A sequence with the property above is said to be subadditive. It can be shown that

$$\frac{a(n)}{n}$$

converges to some limit $\ell(d)$ (see, for instance, Lemma 5.2.1 in Lawler 1991).

(c) Show that

$$d \leq \ell(d) \leq 2d - 1.$$

(Use Problem 7.)

9. So far we have dealt with bond (or edge) percolation: every bond is open with probability p independently of all other bonds. We now consider the slightly different model of site percolation. Again every site is open with probability p independently of all other sites.

(a) Simulate a 50×19 matrix A such that every entry is uniform on $[0, 1]$.

(b) Let p be in $(0, 1)$ and let B be the matrix derived from A in the following way. If $A(i, j) < p$, then set $B(i, j) = 1$. If $A(i, j) > p$, then set $B(i, j) = 0$. Show that the matrix B can be thought of as a simulation of site percolation on $\mathbf{Z}^2$.

(c) Simulate B for $p = 0.1$, $p = 0.5$ and $p = 0.8$. Try to find the biggest open cluster in each case.

(d) Based on your simulations do you think this model exhibits a phase transition?

10. Show that if X is a positive integer-valued random variable then $P(X \geq 1) \leq E(X)$.

2 Further Properties of Percolation

2.1 Continuity of the Percolation Probability

Recall that $\theta(p)$ is the probability that the origin is in an infinite open cluster for the percolation model with parameter p. We have shown that there is a value p_c (that depends on the dimension of the lattice $\mathbf{Z}^d$) in $(0, 1)$ such that $\theta(p) = 0$ for $p < p_c$ and $\theta(p) > 0$ for $p > p_c$.

What happens at p_c? This has been an open question for more than 50 years. It is known that $\theta(p_c) = 0$ in $d = 2$ and for large d but the question is open for $d = 3$, for instance. This is an important question since its answer would tell us whether the phase transition is continuous or discontinuous.

As the next result shows right continuity is not an issue.

Proposition 2.1. *The function θ is right continuous on $[0,1]$.*

As we see next left continuity at p_c is the open question.

Proposition 2.2. *The function θ is left continuous on $[0,1]$ except possibly at p_c.*

By Propositions 2.1 and 2.2 θ is continuous everywhere except possibly at p_c. Unfortunately this is exactly the point we are really interested in!

Proof of Proposition 2.1. We need two lemmas from analysis. First a definition.

Definition 2.1. A real valued function g is said to be upper semicontinuous at t if for each constant c such that $c > g(t)$ there is a $\delta > 0$ such that if $|h| < \delta$ then $c \geq g(t + h)$.

Lemma 2.1. *Assume that the functions g_i are continuous at t. Then $g = \inf_i g_i$ is upper semicontinuous at t.*

Proof of Lemma 2.1. Assume $c > g(t)$ then by definition of g there is i_0 such that $c > g_{i_0}(t) \geq g(t)$. But g_{i_0} is continuous so for any $\epsilon > 0$ there is $\delta > 0$ such that if $|h| < \delta$ then

$$g_{i_0}(t + h) \leq g_{i_0}(t) + \epsilon.$$

Pick $\epsilon = (c - g_{i_0}(t))/2 > 0$ to get

$$g_{i_0}(t + h) \leq c.$$

This implies that

$$g(t + h) \leq c.$$

This concludes the proof that g is upper semicontinuous at t.

We also need the following.

Lemma 2.2. *If g is upper semicontinuous and increasing, it is right continuous.*

Proof of Lemma 2.2. Fix $t > 0$, let $c = g(t) + \epsilon$, by semicontinuity there is $\delta > 0$ such that if $|h| < \delta$ then

$$g(t + h) \leq c = g(t) + \epsilon.$$

Now take h in $(0, \delta)$. We use that g is an increasing function to get

$$0 \leq g(t + h) - g(t) \leq \epsilon.$$

and this shows that g is right continuous. This concludes the proof of Lemma 2.2.

We are now ready for the proof of Proposition 2.1.

Let $S(n)$ be the boundary of the square with side length $2n$. That is,

$$S(n) = \{(x_1, x_2) \in \mathbf{Z}^2 : |x_1| = n \text{ or } |x_2| = n\}.$$

Let $\{0 \longrightarrow S(n)\}$ be the event that there is a path of open edges from the origin to some point in $S(n)$. Note that the sequence of events $\{0 \longrightarrow S(n)\}$ is decreasing in n. That is, if there is an open path from the origin to $S(n + 1)$, then there must be an open path from the origin to $S(n)$ (why?).

By Proposition 1.1 in the Appendix

$$\lim_{n \to \infty} P(0 \longrightarrow S(n)) = P(\bigcap_{n \geq 1} \{0 \longrightarrow S(n)\}).$$

But the event $\cap_{n \geq 1} \{0 \longrightarrow S(n)\}$ happens if and only if the origin is in an infinite open cluster. Therefore, if we let $g_n(p) = P(0 \longrightarrow S(n))$, we have

$$\theta(p) = \lim_{n \to \infty} g_n(p) = \inf_n g_n(p)$$

where the last equality comes from the fact the sequence $(g_n(p))_{n \geq 1}$ is decreasing for any fixed p. Note that the event $\{0 \longrightarrow S(n)\}$ depends only on the edges inside the finite box $B(n)$ where

$$B(n) = \{(x_1, x_2) \in \mathbf{Z}^2 : |x_1| \leq n \text{ and } |x_2| \leq n\}.$$

That is, if we open or close edges outside $B(n)$ it does not affect whether $\{0 \longrightarrow S(n)\}$ happens (why?). Using that there are finitely many edges in $B(n)$ it is not difficult to show that g_n is continuous on $[0, 1]$, see the problems.

For fixed p, $\theta(p)$ is the limit of $g_n(p)$ as n goes to infinity. However, a pointwise limit of continuous functions is not necessarily continuous (see the problems). We can only say (by Lemma 2.1) that θ is upper semicontinuous. Since θ is also increasing (Proposition 1.1) it is right continuous (by Lemma 2.2) on [0,1]. This completes the proof of Proposition 2.1.

A self-contained proof of Proposition 2.2 is beyond the scope of this book. We will use without proving it the following important result.

Theorem 2.1. *There is at most one infinite open cluster for the percolation model on* $\mathbf{Z}^d$.

For a proof of Theorem 2.1 see Bollobas and Riordan (2006) or Grimmett (1999). Putting together Theorem 2.1 and Proposition 1.2 we see that with probability one there is a unique infinite open cluster when $p > p_c$.

Proof of Proposition 2.2. Recall the following construction from the previous section. For each edge e in $\mathbf{Z}^d$ let $U(e)$ be a uniform random variable on [0,1]. The $U(e)$ are mutually independent. We declare an edge e to be open for the percolation model with parameter p if and only if $U(e) < p$. This allows the simultaneous construction of percolation models for all $p \in [0, 1]$.

Let C_p be the open cluster of the origin for the percolation model with parameter p. According to the construction we know that

$$\text{if } p_1 < p_2 \text{ then } C_{p_1} \subset C_{p_2}.$$

Fix $p > p_c$. We want to show that θ is left continuous at p. That is, we want to show that the limit from the left $\lim_{t \to p^-} \theta(t)$ is $\theta(p)$. Note that

$$\lim_{t \to p^-} \theta(t) = \lim_{t \to p^-} P(|C_t| = \infty) = P(|C_t| = \infty \text{ for some } t < p),$$

where we are using the fact that the events $\{|C_t| = \infty\}$ are increasing in t.

Since $C_t \subset C_p$ for all $t < p$ we have

$$0 \le \theta(p) - \lim_{t \to p^-} \theta(t) = P(|C_p| = \infty) - P(|C_t| = \infty \text{ for some } t < p)$$

$$= P(|C_p| = \infty; |C_t| < \infty \text{ for all } t < p)$$

where the last equality comes from the elementary fact that if $A \subset B$ then $P(B) - P(A) = P(B \cap A^c)$. We now have to show that

$$P(|C_p| = \infty; |C_t| < \infty \text{ for all } t < p) = 0.$$

Assume that $|C_p| = \infty$. Let $p_c < t_0 < p$. It follows from Proposition 1.2 that with probability one there is somewhere an infinite cluster for the model with parameter t_0. We denote this infinite cluster by B_{t_0}. Moreover, B_{t_0} must intersect C_p, otherwise we would have two infinite clusters for the model with parameter p (recall that $t_0 < p$), contradicting Theorem 2.1. Thus, there must exist an open path γ from the origin to some point in B_{t_0} for the percolation model with parameter p. This open path is finite and for every edge in the path we must have $U(e)$ strictly less than p

(i.e., every edge is open). Let t_1 be the maximum of the $U(e)$ for all the edges in γ. This maximum is necessarily strictly less than p (why?). Now take $t_2 = \max(t_0, t_1)$. The path γ is open for the model with parameter t_2 and the infinite open cluster B_{t_0} is also open for the percolation model with parameter t_2. Thus, the origin is in an infinite cluster for the model with parameter t_2. In other words $|C_{t_2}| = \infty$ for some $t_2 < p$. This shows that

$$P(|C_p| = \infty; |C_t| < \infty \text{ for all } t < p) = 0.$$

This proves that θ is left continuous on $(p_c, 1]$. By definition θ is identically 0 on $[0, p_c)$ so it is continuous on $[0, p_c)$. This concludes the proof of Proposition 2.2.

2.2   The Subcritical Phase

We have seen that if $p > p_c$ then there is a unique infinite open cluster with probability one. If $p < p_c$ we know that all clusters are finite. What else can we say about this subcritical phase? Recall that $S(n)$ is the boundary of the square with side length $2n$:

$$S(n) = \{(x_1, x_2) \in \mathbf{Z}^2 : |x_1| = n \text{ or } |x_2| = n\}.$$

Let $\{0 \longrightarrow S(n)\}$ be the event that there is a path of open edges from the origin to some point in $S(n)$. The next theorem shows that the probability that the origin be connected to $S(n)$ decays exponentially in n if $p < p_c$.

Theorem 2.2. *Assume that $p < p_c$. Then there is $\alpha > 0$ (depending on p) such that for all $n \geq 1$*

$$P(0 \longrightarrow S(n)) \leq e^{-\alpha n}.$$

Our proof of Theorem 2.2 will not be self-contained. We will need two important results that we will not prove.

First we will need a result relating the expected size of the open cluster containing the origin to the critical value p_c. Define another critical value by

$$p_T = \inf\{p : E(|C|) = \infty\}.$$

Do we have that $p_c = p_T$? This question was open for a long time and is known as the question of uniqueness of the critical point. It is easy to see that $p_c \geq p_T$ (see the problems). What is much less clear is whether $p_c = p_T$. The answer is yes but the proof is fairly involved so we will skip it. See Grimmett (1999) for a proof.

Second we will need a version of the van den Berg–Kesten (BK) inequality. Let A be the event that there is an open path joining vertex x to vertex y and let B be the event that there is an open path joining vertex u to vertex v. Define $A \circ B$ as the event that there are two disjoint open paths, the first joining x to y the second joining u to v. Note that $A \circ B \subset A \cap B$. The BK inequality states that

$$P(A \circ B) \leq P(A)P(B).$$

To give some intuition on the BK inequality, consider the conditional probability $P(A \circ B | A)$. That is, given that there is at least one open path from x to y what is the probability that there is an open path from u to v that uses none of the open edges of one of the paths linking x to y. Intuitively, this condition should make the occurrence of B more difficult. That is, we should have

$$P(A \circ B | A) \leq P(B).$$

This is precisely what the BK inequality states. For a proof of the BK inequality see Grimmett (1999). We will now give a proof of Theorem 2.2 that uses the equality $p_c = p_T$ and the BK inequality.

Proof of Theorem 2.2. Let $S(x,k)$ be the square with side $2k$ and center x. In other words,

$$S(x,k) = \{y \in \mathbf{Z}^2 : y = x + z \text{ for some } z \in S(k)\}.$$

Observe that in order to have a path of open edges from 0 to $S(n+k)$ we must have a path of open edges from 0 to some x in $S(n)$ and then a path from x to $S(x,k)$. Moreover, we may take these two paths to be disjoint (why?). Thus,

$$P(0 \longrightarrow S(n+k)) \leq \sum_{x \in S(n)} P(\{0 \longrightarrow x\} \circ \{x \longrightarrow S(x,k)\}).$$

By the BK inequality we get

$$P(0 \longrightarrow S(n+k)) \leq \sum_{x \in S(n)} P(0 \longrightarrow x)P(x \longrightarrow S(x,k)).$$

By translation invariance we have that $P(x \longrightarrow S(x,k)) = P(0 \longrightarrow S(k))$ for any x. Therefore,

$$P(0 \longrightarrow S(n+k)) \leq P(0 \longrightarrow S(k)) \sum_{x \in S(n)} P(0 \longrightarrow x).$$

If we let

$$a(n) = P(0 \longrightarrow S(n))$$

and

$$b(n) = \sum_{x \in S(n)} P(0 \longrightarrow x)$$

then the preceding inequality can be rewritten as

$$a(n + k) \leq a(k)b(n). \tag{2.1}$$

This completes the first step of the proof.

The next step will be to show that $b(n)$ converges to 0 and this together with (2.1) will imply that $a(n)$ goes to 0 exponentially fast.

In order to interpret $b(n)$ we need a little more notation. Let 1_A be the indicator of the event A. If A occurs then the random variable 1_A is 1, if A does not occur then 1_A is 0. In particular the expected value $E(1_A) = P(A)$. Note that

$$\sum_{x \in S(n)} 1_{\{0 \longrightarrow x\}}$$

counts the number of points of $S(n)$ that can be joined to the origin by using an open path. Thus,

$$E(\sum_{x \in S(n)} 1_{\{0 \longrightarrow x\}}) = \sum_{x \in S(n)} P(0 \longrightarrow x) = b(n).$$

Therefore, $b(n)$ is the expected number of points of $S(n)$ that can be joined to the origin by using an open path. Observe now that

$$|C| = \sum_{x \in \mathbf{Z}^d} 1_{\{0 \longrightarrow x\}} = \sum_{n \geq 0} \sum_{x \in S(n)} 1_{\{0 \longrightarrow x\}}$$

where the last equality comes from the fact that the sets $S(n)$ for $n \geq 0$ partition $\mathbf{Z}^d$. Since all terms are positive we may interchange the infinite sum and the expectation to get

$$E(|C|) = \sum_{n \geq 0} b(n).$$

We now use that $p_c = p_T$. Since $p < p_c = p_T$ we have that $E(|C|) < \infty$. Thus,

$$\lim_{n \to \infty} b(n) = 0.$$

This completes the second step of the proof.

In the third and final step of the proof we use (2.1) and the convergence of $b(n)$ to show the exponential convergence of $a(n)$. Since $b(n)$ converges to 0 there is an N such that if $n \geq N$ then $b(n) < 1/2$. Take $n > N$. We have $n = Ns + r$ for some positive integers $r < N$ and s. Since $n \geq Ns$,

$$a(n) = P(0 \longrightarrow S(n)) \leq P(0 \longrightarrow S(Ns)) = a(Ns).$$

Using (2.1) we get

$$a(Ns) \leq a(N(s-1))b_N.$$

Since $b_N < \frac{1}{2}$ we have for every natural s that

$$a(Ns) \leq \frac{1}{2}a(N(s-1)).$$

It is now easy to show by induction on s that

$$a(Ns) \leq (\frac{1}{2})^s a(0) = (\frac{1}{2})^s$$

since $a(0) = 1$. Hence,

$$a(n) \leq (\frac{1}{2})^s = (\frac{1}{2})^{\frac{n-r}{N}} \leq (\frac{1}{2})^{\frac{n}{N}-1} = 2e^{-\frac{\ln 2}{N}n} \text{ for } n \geq N$$

where the second inequality comes from the fact that $r < N$.

At this point we are basically done. We just need to get rid of the "2" in the formula above and of the condition $n \geq N$. Note that for any $0 < \sigma < \ln 2/N$ there is an integer N_1 such that

$$2e^{-\frac{\ln 2}{N}n} < e^{-\sigma n} \text{ for all } n \geq N_1.$$

This is so because the ratio of the l.h.s. over the r.h.s. goes to 0 as n goes to infinity. Let $N_2 = \max(N, N_1)$. We have

$$a(n) \leq e^{-\sigma n} \text{ for all } n \geq N_2.$$

We now take care of $n < N_2$. Note that $a(n) < 1$ for $n \geq 1$. Hence, there is an $\alpha(n) > 0$ such that $a(n) = e^{-\alpha(n)n}$ for $n \geq 1$. Let α be the minimum of $\{\alpha(1), \ldots, \alpha(N_2 - 1), \sigma\}$. Then $\alpha > 0$ (why?) and we have

$$a(n) \leq e^{-\alpha n} \text{ for all } n \geq 1.$$

This completes the proof of Theorem 2.2.

Problems

1. Show that the function θ is continuous at p_c if and only if $\theta(p_c) = 0$.

2. Let

$$B(n) = \{(x_1, x_2) \in \mathbf{Z}^2 : |x_1| \le n \text{ and } |x_2| \le n\}.$$

Let $g_n(p) = P_p(0 \longrightarrow S(n))$ where $S(n)$ is the boundary of $B(n)$. For each edge e in $\mathbf{Z}^2$ let $U(e)$ be uniform in $[0, 1]$. Take the $U(e)$ mutually independent. These variables have been used to construct simultaneously the percolation models for all p in $[0, 1]$, see the proof of Proposition 1.1.

(a) Let $p_1 < p_2$. Show that an edge e is open for the model with p_2 and closed for the model with p_1 if and only if $p_1 < U(e) < p_2$.

(b) Use (a) to show that

$$0 \le g_n(p_2) - g_n(p_1) \le |B(n)|(p_2 - p_1).$$

(c) Use (b) to show that g_n is a continuous function on $[0, 1]$.

3. In this problem we give an example of a sequence of continuous functions that converge pointwise to a discontinuous function.

Let $f_n(p) = 1 - np$ for p in $[0, 1/n]$ and $f_n(p) = 0$ for p in $(1/n, 1]$.

(a) Show that for each $n \ge 1$, f_n is continuous on $[0,1]$.
(b) Show that for each p in $[0,1]$, $\lim_{n\to\infty} f_n(p)$ exists. Denote the limit by $f(p)$.
(c) Show that f is not a continuous function on $[0,1]$.

4. Let

$$a(n) = P(0 \longrightarrow S(n))$$

and

$$b(n) = \sum_{x \in S(n)} P(0 \longrightarrow x).$$

(a) Show that $a(n) < 1$ for every $n \ge 1$.
(b) Show that $a(n) \le b(n)$ for every $n \ge 1$.

5. (a) Prove that for any $n \ge 1$

$$E(|C|) \ge nP(|C| \ge n) \ge nP(|C| = \infty).$$

(b) Prove that if $p > p_c$ then $E(|C|) = \infty$.

(c) Prove that $p_T \le p_c$.

6. Let A be the event that there is an open path joining vertex x to vertex y and let B be the event that there is an open path joining vertex u to vertex v.

(a) Explain why A and B are not independent.

(b) Define $A \circ B$ as the event that there are two disjoint open paths, the first joining x to y the second joining u to v. Show that if $P(A \circ B \mid A) \le P(B)$ then $P(A \circ B) \le P(A)P(B)$.

7. Let

$$S(n) = \{(x_1, x_2) \in \mathbf{Z}^2 : |x_1| = n \text{ or } |x_2| = n\}.$$

Let $S(x, k) = S(k) + x$ for x in $\mathbf{Z}^2$. We pick x so that $S(n)$ and $S(x, k)$ do not intersect. Let

$$A = \{0 \longrightarrow S(n)\}$$

and

$$B = \{x \longrightarrow S(x, k)\}.$$

Show that A and B are independent.

8. Assume that $a(n)$ is a sequence of positive real numbers such that

$$a(n + k) \le a(n)a(k) \text{ and } \lim_{n \to \infty} a(n) = 0.$$

Show that there are real numbers $c > 0$ and $\alpha > 0$ such that

$$a(n) \le ce^{-\alpha n} \text{ for all } n \ge 1.$$

9. A real function g is said to be lower semicontinuous if $-g$ is upper semicontinuous.

(a) State and prove a result analogous to Lemma 2.1 for a lower semi-continuous function.

(b) State and prove a result analogous to Lemma 2.2 for a lower semi-continuous function.

(c) Prove that g is continuous at x if and only if g is upper and lower semi-continuous at x.

3 Percolation On a Tree and Two Critical Exponents

As pointed out in previous sections, critical values such as $p_c(d)$ (the critical value of percolation on the lattice $\mathbf{Z}^d$) are usually unknown (a notable exception is in dimension $d = 2$, $p_c(2) = \frac{1}{2}$). Moreover, it is believed that critical values depend rather heavily on the details of the models making them not that interesting. On the other hand, *critical exponents*, which we introduce below, are more interesting objects since they are believed to be the same for large classes of models. This is known, in mathematical physics, as the universality principle. This principle has been rigorously verified in only a few cases.

The main purpose of this section is to compute two critical exponents that we now introduce.

It is believed that the limit from the right

$$\lim_{p \to p_c^+} \frac{\log \theta(p)}{\log(p - p_c)} = \beta \text{ exists.}$$

In other words, it is believed that near p_c $\theta(p)$ behaves like $(p - p_c)^\beta$. The number β is an example of a critical exponent. It is also believed that the limit from the left

$$\lim_{p \to p_c^-} \frac{\log E|C|}{\log(p_c - p)} = -\gamma \text{ exists}$$

and γ is another critical exponent.

We will now compute these two critical exponents for percolation on a tree. We first describe the rooted binary tree. There is a distinguished site that we call the root. The root has two children and each child has also two children and so on. The root is generation 0, the children of the root are in generation 1 and so on. In general, generation n has 2^n individuals (also called sites). Observe that the root has only two nearest neighbors while all the other sites have three nearest neighbors. So there are three edges at each site except for the root. See Fig. 10.3.

The big advantage of this graph over $\mathbf{Z}^d$ (for $d \geq 2$) is that on the tree there are no loops and exact computations are possible.

Let p be in $[0, 1]$. We declare each edge open with probability p and closed with probability $1 - p$ independently of all the other edges. Let C be the set of sites that

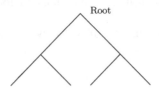

Fig. 10.3 This is part of the infinite rooted tree

are linked to the root by an open path in the tree. In other words, C is the open cluster containing the root. It turns out that there is a second interpretation for C that is very useful. Define $Z_0 = 1$. For a given parent, the number Y of children that are linked to the parent by an open edge is a binomial $(2, p)$ random variable. That is,

$$P(Y = k) = \binom{2}{k} p^k (1 - p)^{2-k} \text{ for } k = 0, 1, 2.$$

Let Z_n be the number of individuals in the nth generation that are connected to the root by an open path. Then we have for $n \geq 1$

$$Z_n = \sum_{i=1}^{Z_{n-1}} Y_i$$

where the Y_i are i.i.d. random variables with a binomial $(2, p)$ distribution. Thus, Z_n is a Bienaymé–Galton–Watson process. A little thought reveals that the relation between Z_n and C is given by

$$|C| = \sum_{n \geq 0} Z_n.$$

From this we see that

$$\theta(p) = P(|C| = \infty) = P(Z_n \geq 1, \text{ for all } n \geq 1).$$

We know that a BGW survives if and only if the mean offspring is strictly larger than 1. Hence, $\theta(p) > 0$ if and only if $E(Y) = 2p > 1$. Thus, the critical parameter $p_c = \frac{1}{2}$ for the percolation model on the binary tree.

If $p > 1/2$, the extinction probability is the unique solution smaller than one of the equation $f(s) = s$ where f is the generating function for Y. We have

$$f(s) = p^2 s^2 + 2p(1 - p)s + (1 - p)^2.$$

We solve

$$f(s) = s$$

and get two roots: $s = 1$ and $s = \frac{(1-p)^2}{p^2}$. So the survival probability is

$$\theta(p) = 1 - \frac{(1 - p)^2}{p^2} \text{ for } p > 1/2.$$

A little algebra gives

$$\theta(p) = \frac{2p-1}{p^2}.$$

We may compute the limit from the right

$$\lim_{p \to 1/2+} \frac{\ln(\theta(p))}{\ln(p-1/2)} = 1.$$

This shows that the critical exponent β exists for percolation on this tree and that it is equal to 1.

We now turn to the computation of $E|C|$. We know that for a BGW process

$$E(Z_n) = (EY)^n = (2p)^n.$$

Thus, for $p < 1/2$

$$E|C| = E(\sum_{n \geq 0} Z_n) = \sum_{n \geq 0} (2p)^n = \frac{1}{1-2p}.$$

Thus, the limit from the left is

$$\lim_{p \to p_c^-} \frac{\log E|C|}{\log(p_c - p)} = -1,$$

and this shows that the critical exponent γ exists and that it is also equal to 1.

Problems

1. Consider percolation on a rooted tree where each site has three children instead of two.

(a) Compute the critical value p_c.
(b) Show that the critical exponents are the same as for two children.
(c) Can you guess what happens for percolation on a tree with d children?

2. Explain why the relation

$$|C| = \sum_{n \geq 0} Z_n$$

is true.

3. Consider percolation on the binary tree. Show that

(a)

$$\lim_{p \to 1/2+} \frac{\ln(\theta(p))}{\ln(p - 1/2)} = 1.$$

(b)

$$\lim_{p \to 1/2-} \frac{\log E|C|}{\log(1/2 - p)} = -1.$$

4. (a) Show that the function θ is continuous on $[0, 1]$ for the percolation model on the tree.

(b) Is the function θ differentiable at p_c?

5. Assume that $p < p_c$. Show that there exists a in $(0, 1)$ such that

$$P(|C| \geq n) \leq a^n.$$

6. (a) Assume that there is some ℓ such that

$$\lim_{x \to 1} \frac{f(x)}{(x - 1)^2} = \ell.$$

Compute

$$\lim_{x \to 1+} \frac{\ln(f(x))}{\ln((x - 1)^2)}.$$

(b) Give an example of f for which $\lim_{x \to 1+} \frac{\ln(f(x))}{\ln((x-1)^2)}$ exists but $\lim_{x \to 1} \frac{f(x)}{(x-1)^2}$ does not.

Notes

This chapter is a small introduction to the subject. We have mentioned some important questions. Is there a phase transition? How many infinite open clusters are there? Is there percolation at $p_c(d)$? What is the typical size of a cluster in the subcritical region? We only proved a small part of the corresponding results. Grimmett (1999) gives a self-contained exposition of percolation theory. See also Bollobas and Riordan (2006). Percolation on the tree is somehow trivial since it is equivalent to a branching model. However, it suggests conjectures on more intricate graphs.

References

Bollobas, B., Riordan, O.: Percolation. Cambridge University Press, Cambridge (2006)
Durrett, R.: Probability Theory and Examples, 4th edn. Cambridge University Press, Cambridge (2010)
Grimmett, G.: Percolation. Springer, New York (1999)
Lawler, G.F.: Intersections of Random Walks. Birkhauser, Boston (1991)

Chapter 11
A Cellular Automaton

Cellular automata are widely used in mathematical physics and in theoretical biology. These systems start from a random state and evolve using deterministic rules. We concentrate on a specific model in this chapter. The techniques we use are similar to the ones used in percolation.

1 The Model

We define a cellular automaton on $\mathbf{Z}^2$. Each site of $\mathbf{Z}^2$ is in state 1 or 0. Let η_n be the state (or configuration) of the system at time n where n is a positive integer. Let x be in $\mathbf{Z}^2$. We let $\eta_n(x) = 1$ if x is occupied at time n. We let $\eta_n(x) = 0$ if x is empty at time n.

The initial state for the cellular automaton is the following. Let p be in [0,1]. For each site x of $\mathbf{Z}^2$ we put a 1 with probability p and a 0 with probability $1 - p$. This is done independently for each site. Thus, for each x in $\mathbf{Z}^2$

$$P(\eta_0(x) = 1) = p \qquad P(\eta_0(x) = 0) = 1 - p.$$

The system evolves according to the following rules. If $\eta_n(x) = 1$, then $\eta_{n+1}(x) = 1$. If $\eta_n(x) = 0$ and if at least one neighbor in each of the orthogonal directions is a 1, then $\eta_{n+1}(x) = 1$. In words, once a site is occupied it remains occupied forever. An empty site becomes occupied if it has at least two (among four) occupied nearest neighbors in two orthogonal directions. We see that the initial state is random but that the evolution is deterministic. We now give an example to illustrate the rules of evolution of this system. We consider a part of $\mathbf{Z}^2$ where we have the following configuration at time 0:

© Springer Science+Business Media New York 2014
R.B. Schinazi, *Classical and Spatial Stochastic Processes: With Applications to Biology*, DOI 10.1007/978-1-4939-1869-0_11

$$1\ 1\ 1\ 1\ 1$$
$$1\ 0\ 0\ 0\ 1$$
$$1\ 0\ 0\ 0\ 1$$
$$1\ 1\ 1\ 1\ 1$$

Then at time 1 we have in the same region

$$1\ 1\ 1\ 1\ 1$$
$$1\ 1\ 0\ 1\ 1$$
$$1\ 1\ 0\ 1\ 1$$
$$1\ 1\ 1\ 1\ 1$$

and at time 2 we have

$$1\ 1\ 1\ 1\ 1$$
$$1\ 1\ 1\ 1\ 1$$
$$1\ 1\ 1\ 1\ 1$$
$$1\ 1\ 1\ 1\ 1$$

Define

$$A_n = \{\eta_n(0) = 1\}$$

where 0 is the origin of $\mathbf{Z}^2$. Since a 1 does not change its state, $A_n \subset A_{n+1}$. So $P(A_n)$ is an increasing sequence in [0,1] and therefore, converges. We define

$$\lim_n P(A_n) = \rho(p).$$

By Proposition 1.1 in the Appendix we get

$$\rho(p) = P(\bigcup_{n \geq 1} A_n).$$

That is, $\rho(p)$ is the probability that the origin will eventually be occupied. It is natural to define the critical value

$$p_c = \inf\{p \in [0, 1] : \rho(p) = 1\}.$$

This model is translation invariant so $\rho(p)$ is the same for all the sites of $\mathbf{Z}^2$. In particular if $\rho(p) = 1$ it means that all the sites of $\mathbf{Z}^2$ will eventually be occupied. It is actually possible to compute p_c.

Theorem 1.1. *The cellular automaton on $\mathbf{Z}^2$ has $p_c = 0$.*

Theorem 1.1 tells us that the cellular automaton started with any density $p > 0$ will fill the whole space. This result is far from obvious. If p is very small there are very few 1's at the beginning and it is not clear that the whole space is going to get filled by 1's. The following proof gives some intuition on how the space gets filled.

Proof of Theorem 1.1. We say that site x is a good site if for each integer k larger than 1 the square whose center is x and whose sides have length $2k$ has at least one 1 on each of its four sides (excluding the four vertices) at time 0. An induction proof on k shows that if x is a good site then each square whose center is x will eventually be filled by 1's. Thus, if x is a "good site" the whole space will eventually be filled. Let S_k be the square whose center is the origin of $\mathbf{Z}^2$ and whose sides have length $2k$. For $k \geq 1$ let $B_k = \{$ each side of S_k has at least one 1 at time 0 that is not on one of the vertices of the square $\}$. Let B_0 be the event that the origin is occupied at time 0.

Since each side of S_k has $2k - 1$ sites that are not vertices of the square we get for $k \geq 1$,

$$P(B_k) = (1 - q^{2k-1})^4$$

where $q = 1 - p$. Let $C_k = \bigcap_{n=0}^{k} B_n$. Note that $(C_k)_{k \geq 0}$ is a decreasing sequence of events, thus

$$\lim_{k \to \infty} P(C_k) = P(\bigcap_{k=0}^{\infty} C_k).$$

Let

$$E = \bigcap_{k \geq 0} B_k.$$

Observe that $E = \{$ the origin is a good site $\}$. Note that E can also be written as

$$E = \bigcap_{k \geq 0} C_k.$$

Since the B_n are independent events (why?) we have

$$P(C_k) = p(1 - q)^4(1 - q^3)^4 \ldots (1 - q^{2k-1})^4.$$

Hence,

$$P(E) = \lim_{k \to \infty} p(1 - q)^4(1 - q^3)^4 \ldots (1 - q^{2k-1})^4.$$

That is, $P(E)$ is the infinite product

$$\Pi_{k=1}^{\infty}(1 - q^{2k-1})^4.$$

Observe that the series

$$\sum_{k \geq 1} q^{2k-1}$$

converges for every $p > 0$. This implies that the infinite product $\Pi_{k=1}^{\infty}(1 - q^{2k-1})^4$ is strictly positive (see the Appendix). So we have

$$P(0 \text{ is a good site}) = P(E) > 0.$$

Consider the event

$$F = \bigcup_{x \in \mathbf{Z}^2} \{x \text{ is a good site}\}.$$

That is, F is the event that there is a good site somewhere in $\mathbf{Z}^2$. This is a translation invariant event that depends only on the initial distribution which is a product measure. By the ergodic Theorem (see Durrett 2010) $P(F) = 0$ or 1. Since $P(F) \geq P(E) > 0$ we have that $P(F) = 1$. Thus, $\mathbf{Z}^2$ will eventually be filled by 1's for any $p > 0$. This shows that $p_c = 0$ and completes the proof of Theorem 1.1.

Problems

1. Consider the cellular automaton in a finite subset of $\mathbf{Z}^2$. Show that $p_c = 1$.

2. Consider a cellular automaton on $\mathbf{Z}$ for which a 0 becomes a 1 if one of its two nearest neighbors is a 1.

(a) Show that

$$\{\eta_n(0) = 0\} = \{\eta_0(x) = 0 \text{ for } x = -n, \dots, n\}.$$

(b) Show that for any $p > 0$

$$\lim_{n \to \infty} P(\eta_n(0) = 0) = 0.$$

(c) Show that $p_c = 0$.

3. Consider a cellular automaton on $\mathbf{Z}^2$ with a different rule. This time a 0 becomes a 1 if at least two of its four nearest neighbors are occupied by 1's. Let ξ_n be the configuration of this automaton at time n and let η_n be the automaton we have considered so far.

(a) Show that we can couple the two automata in such a way that for every x in $\mathbf{Z}^2$ and every $n \geq 0$ we have

$$\eta_n(x) \leq \xi_n(x).$$

(b) Use (a) to show that the p_c corresponding to ξ_n is also 0.

4. Consider a cellular automaton on $\mathbf{Z}^2$ for which a 0 becomes a 1 if at least three of its four nearest neighbors are occupied by 1's.

(a) Assume that the 3×3 square centered at the origin of $\mathbf{Z}^2$ is empty. That is, assume that the initial configuration is

$$0\ 0\ 0$$
$$0\ 0\ 0$$
$$0\ 0\ 0$$

Explain why all the sites of the square will remain empty forever.

(b) Show that $p_c = 1$.

5. Let

$$A_n = \{\eta_n(0) = 1\}.$$

Show that $A_n \subset A_{n+1}$.

6. Let S_k be the square whose center is the origin of $\mathbf{Z}^2$ and whose sides have length $2k$. For $k \geq 1$ let

$B_k = \{$ each side of S_k has at least one 1 at time 0 that is not on one of the vertices of the square $\}$.

Let B_0 be the event that the origin is occupied at time 0.

(a) Draw a configuration for which B_0, B_1, B_2 and B_3 hold.

(b) Show that for $k \geq 1$,

$$P(B_k) = (1 - q^{2k-1})^4$$

where $q = 1 - p$.

(c) Let $C_k = \bigcap_{n=0}^{k} B_n$. Show that $(C_k)_{k \geq 0}$ is a decreasing sequence of events.

(d) Show that $\bigcap_{n=0}^{k} C_n = \bigcap_{n=0}^{k} B_n$.

2   A Renormalization Argument

Let T be the first time the origin is occupied:

$$T = \inf\{n \geq 0 : \eta_n(0) = 1\}.$$

Note that

$$P(T > n) = P(\eta_n(0) = 0).$$

Theorem 1.1 shows that for any $p > 0$ we have

$$\lim_{n \to \infty} P(\eta_n(0) = 1) = 1.$$

Therefore,

$$\lim_{n \to \infty} P(\eta_n(0) = 0) = \lim_{n \to \infty} P(T > n) = 0.$$

The sequence of events $\{T > n\}$ is decreasing. Hence,

$$\lim_{n \to \infty} P(T > n) = P(\bigcap_{n=0}^{\infty} \{T > n\}) = 0.$$

Observe that $T > n$ for all n if and only if $T = +\infty$ (i.e., the origin stays empty forever). Hence,

$$P(T = +\infty) = 0.$$

Thus, with probability one there is an n such that the origin is occupied after time n. In other words, the random time T is finite with probability one. The main purpose of this section is to prove that T decays exponentially for any $p > 0$ in dimension 2. We will proceed in two steps. First we will prove that T decays exponentially for $p > 2/3$. The second step will be to show that when p is in $(0, 2/3)$ we may rescale time and space so that the rescaled system is a cellular automaton with $p > 2/3$. This will prove the exponential decay for all $p > 0$.

Theorem 2.1. *For any $p > 0$ there are $\gamma(p)$ and $C(p)$ in $(0, \infty)$ such that*

$$P(T \geq n) \leq C(p)e^{-\gamma(p)n}.$$

Proof of Theorem 2.1.
   First step
   We prove the Theorem for $p > 2/3$. We will use a technique we have used in percolation.

Let $C_n(0)$ be the empty cluster of the origin at time n. That is, x is in $C_n(0)$ if there is an integer $k \geq 1$ and if there is a sequence of sites $x_0, x_1, \ldots, x_k$ in $\mathbf{Z}^2$ such that $x_0 = 0$, $x_k = x$, x_i and x_{i+1} are nearest neighbors for $i = 0, \ldots, k - 1$, and $\eta_n(x_i) = 0$ for $i = 1, \ldots, k$. Let

$$D_k = \{(m,n) \in \mathbf{Z}^2 : |m| + |n| = k\}$$

and let

$$R_n(0) = \sup\{k : C_n(0) \cap D_k \neq \emptyset\}.$$

That is, $R_n(0)$ plays the role of a radius for the empty cluster. We first show that $R_0(0)$ is finite. For $R_0(0) \geq n$ there must be at time 0 a self-avoiding walk of empty sites that starts at the origin and reaches some site in D_n. There are at most $4(3)^{l-1}$ self-avoiding walks with l sites and to reach D_n we need $l \geq n + 1$. Thus,

$$P(R_0(0) \geq n) \leq \sum_{l=n+1}^{\infty} 4(3)^{l-1}(1-p)^l \tag{2.1}$$

The geometric series is convergent if and only if $p > 2/3$. Hence, if $p > 2/3$ we have

$$\lim_{n \to \infty} P(R_0(0) \geq n) = 0.$$

Since the sequence of events $\{R_0(0) \geq n\}$ is decreasing we have

$$\lim_{n \to \infty} P(R_0(0) \geq n) = P(\bigcap_{n \geq 1}\{R_0(0) \geq n\} = P(R_0(0) = +\infty) = 0.$$

Hence, the empty cluster of the origin is finite at time 0. Next we show that this cluster decreases with time. Observe that if $R_n(0) = k$ then every site in $C_n(0) \cap D_k$ will have at least two occupied nearest neighbors, one in each direction (why?) and hence they will become occupied at time $n + 1$. Thus, $R_{n+1}(0) \leq R_n(0) - 1$. That is the radius of the empty cluster will decrease by at least 1 from time n to time $n + 1$. In order for the origin to be empty at time n we therefore need the radius of the empty cluster to be at least n at time 0. That is,

$$P(T > n) \leq P(R_0(0) \geq n).$$

Summing the geometric series in (2.1) yields for $p > 2/3$

$$P(R_0(0) \geq n) \leq 4(1-p)(3(1-p))^n \frac{1}{1 - 3(1-p)} = C(p)e^{-\gamma(p)n}$$

where

$$C(p) = 4(1-p)\frac{1}{1-3(1-p)} \qquad \gamma(p) = -\ln((3(1-p))).$$

That is, if $p > 2/3$ $P(T > n)$ decays exponentially with n. This completes the first step of the proof.

Second step

To prove Theorem 2.1 for $p < 2/3$ we first need some new notation. Recall that S_N is the square centered at the origin whose sides have length $2N$. Assume we start with 1's with probability p and 0's with probability $1 - p$ inside S_N and all the sites outside S_N empty. We do not allow creation of 1's outside the square S_N. The rules inside S_N are the same as before. If all the sites of S_N eventually become occupied with the above restrictions, then S_N is said to be internally spanned. Let $R(N, p)$ be the probability that S_N is internally spanned.

Lemma 2.1. *For any* $0 < p \le 1$ *we have*

$$\lim_{N\to\infty} R(N, p) = 1.$$

Proof of Lemma 2.1. Let M be the largest integer below $N/4$ and let

$$E_N = \{\text{there is a good site in } S_M\}$$

where the definition of a good site has been given in the proof of Theorem 1.1. Observe that E_N is an increasing sequence of events and that

$$\bigcup_{N\ge1} E_N = \bigcup_{x\in\mathbb{Z}^2} \{x \text{ is a good site}\}.$$

The event on the r.h.s. has been shown to have probability 1 in the proof of Theorem 1.1. Thus,

$$\lim_{N\to\infty} P(E_N) = P(\bigcup_{N\ge1} E_N) = 1.$$

Let F_N be the event that for all $k \ge M$ each side of the square S_k has at least one occupied site that is not a vertex. Hence,

$$P(F_N) = \Pi_{i=M}^{\infty}(1 - q^{2i-1})^4$$

and since

$$\sum_{i=1}^{\infty} -\ln(1 - q^{2i-1}) < \infty$$

we have

$$\lim_{N \to \infty} P(F_N) = 1.$$

See the problems.

Observe now that if E_N occurs then S_M will eventually be fully occupied in the system restricted to S_N (why?). If F_N and E_N occur simultaneously, then S_N is internally spanned. Thus,

$$R(N, p) \geq P(E_N \cap F_N)$$

and

$$\lim_{N \to \infty} R(N, p) = 1.$$

This completes the proof of Lemma 2.1.

We are now ready to introduce the renormalization scheme. We start with the renormalized sites. Let S_N be the square whose sides have length $2N$ and that is centered at the origin of $\mathbf{Z}^2$. And for k in $\mathbf{Z}^2$, let the renormalized site k be

$$S_N(k) = \{x \in \mathbf{Z}^2 : x - k(2N + 1) \in S_N\}.$$

The squares $S_N(k)$ cover $\mathbf{Z}^2$.

We say that the renormalized site k is occupied if all the sites in $S_N(k)$ are occupied. Observe that if the renormalized site k has at least one occupied renormalized neighbor in each of the orthogonal directions and if k is not occupied at time n then it will be occupied at time $n + 2N + 1$. Observe also that if $S_N(k)$ is internally spanned it must be so by time $|S_N(k)| = (2N + 1)^2$. This is so because if by time $(2N + 1)^2$ the renormalized site is not internally spanned it means that there is a time $n < (2N + 1)^2$ at which no change happened in the square and after that time there can be no change. The two preceding observations motivate the definition of the renormalized time τ by the following equation.

$$n = (2N + 1)^2 + (2N + 1)\tau.$$

We now describe the dynamics of the renormalized system. For each $S_N(k)$ we do not allow creation of 1's from inside to outside of $S_N(k)$ up to time $(2N + 1)^2$. At time $n = (2N + 1)^2$ we keep the 1's that are in renormalized sites $S_N(k)$ that are fully occupied and we replace all the 1's by 0's in the $S_N(k)$ that are not fully occupied. Thus, at time $\tau = 0$ $(n = (2N + 1)^2))$ each renormalized site is occupied with probability $R(N, p)$ independently of each other. After time $\tau = 0$, the restriction is dropped and 1's may be created from one renormalized site into another. The crucial remark is that the system with renormalized time and space behaves exactly like a cellular automaton on $\mathbf{Z}^2$ with initial density $R(N, p)$. To be more precise, if we identify $S_N(k)$ and k and we look at the system at times $\tau = 0, 1, \ldots$ then the renormalized system is a cellular automaton on $\mathbf{Z}^2$ with initial density $R(N, p)$.

The preceding construction couples the cellular automaton and the renormalized system in such a way that if the origin is empty for the cellular automaton at time n then the renormalized origin is also empty at the corresponding rescaled time

$$\tau = [\frac{n - (2N + 1)^2}{(2N + 1)}]$$

where $[x]$ is the integer part of x. This is so because of the restrictions we impose on the creation of 1's for the renormalized system up to time $(2N + 1)^2$. Thus,

$$P_p(T > n) \le P_{R(N,p)}(T > [\frac{n - (2N + 1)^2}{(2N + 1)}])$$

where the subscripts indicate the initial densities of the two cellular automata we are comparing. By Lemma 2.1, for any $p > 0$ there is N such that

$$R(N, p) > 2/3.$$

We pick such an N. We proved Theorem 2.1 for $p > 2/3$ and since $R(N, p) > 2/3$ we have the existence of γ and C (depending on p and N) such that

$$P_p(T > n) \le P_{R(N,p)}(T > [\frac{n - (2N + 1)^2}{(2N + 1)}]) \le C e^{-\gamma [\frac{n - (2N+1)^2}{(2N+1)}]}.$$

This completes the proof of Theorem 2.1.

Problems

1. Let

$$s_n = \Pi_{i=1}^{n}(1 - q^{2i-1})^4$$

In the proof of Theorem 1.1 we have shown that s_n has a strictly positive limit. Use this fact to prove that for $0 \le q < 1$

$$\lim_{M \to \infty} \Pi_{i=M}^{\infty}(1 - q^{2i-1})^4 = 1.$$

2. Show that if

$$\lim_{N\to\infty} P(F_N) = 1 \text{ and if } \lim_{N\to\infty} P(E_N) = 1$$

then

$$\lim_{N\to\infty} P(E_N \cap F_N) = 1.$$

3. Show that Theorem 2.1 holds for the cellular automaton in dimension 1.

4. Consider the following initial configuration

```
1 0 0 0 1
1 0 0 0 1
1 1 0 0 1
1 1 0 0 1
1 1 1 1 1
```

The origin is at the center of the matrix (third row, third column). For the notation see the proof of Theorem 1.2.

(a) What is $R_0(0)$ in this case?

(b) What is $R_1(0)$?

(c) How long does it take for the origin to be occupied?

5. See the definitions of E_N and F_N in the proof of Theorem 1.2. Set $N = 8$ and $M = 2$.

(a) Give an initial configuration for which E_N and F_N occur.

(b) Show on your example that if E_N occurs then S_M will eventually be fully occupied in the system restricted to S_N.

(c) Show on your example that if F_N and E_N occur then S_N is internally spanned.

Notes

We have followed the treatment of Schonmann (1992). Using the renormalization argument Schonmann goes on to prove that $p_c = 0$ and that convergence to the all 1's state occurs exponentially fast in any dimension.

References

Durrett, R.: Probability Theory and Examples, 4th edn. Cambridge University Press, Cambridge (2010)

Schonmann, R.: On the behavior of some cellular automata related to bootstrap percolation. Ann. Probab. **20**, 174–193 (1992)

Chapter 12
A Branching Random Walk

In this chapter we consider a continuous time spatial branching process. Births and deaths are as in the binary branching process. In addition we keep track of the spatial location of the particles. We use results about the binary branching process.

1 The Model

Consider a system of particles which undergo branching and random motion on some countable set S. Our two main examples of S will be the lattice $\mathbf{Z}^d$ and the homogeneous tree. Let $p(x, y)$ be the probability transitions of a given Markov chain on S. Hence, for every x we have

$$\sum_{y \in S} p(x, y) = 1.$$

The evolution of a continuous time branching random walk on S is governed by the two following rules.

1. Let $\lambda > 0$ be a parameter. If $p(x, y) > 0$, and if there is a particle at x then this particle waits a random exponential time with rate $\lambda p(x, y)$ and gives birth to a new particle at y.
2. A particle waits an exponential time with rate 1 and then dies.

Let $b_t^{x,\lambda}$ denote the branching random walk starting from a single particle at x and let $b_t^{x,\lambda}(y)$ be the number of particles at site y at time t.

© Springer Science+Business Media New York 2014
R.B. Schinazi, *Classical and Spatial Stochastic Processes: With Applications to Biology*, DOI 10.1007/978-1-4939-1869-0_12

We denote the total number of particles of $b_t^{x,\lambda}$ by $|b_t^{x,\lambda}| = \sum_{y \in S} b_t^{x,\lambda}(y)$. Let O be a fixed site of S. We define the following critical parameters.

$$\lambda_1 = \inf\{\lambda : P(|b_t^{O,\lambda}| \geq 1, \forall t > 0) > 0\}$$

$$\lambda_2 = \inf\{\lambda : P(\limsup_{t \to \infty} b_t^{O,\lambda}(O) \geq 1) > 0\}.$$

Note that the event $\{\limsup_{t \to \infty} b_t^{O,\lambda}(O) \geq 1\}$ is the event that there will be at least one particle at O at arbitrarily large times.

In words, λ_1 is the critical value corresponding to the *global* survival of the branching Markov chain while λ_2 is the critical value corresponding to the *local* survival of the branching Markov chain.

Note that

$$\{\limsup_{t \to \infty} b_t^{O,\lambda}(O) \geq 1\} \subset \{|b_t^{O,\lambda}| \geq 1, \forall t > 0\}.$$

That is, if the process survives locally it survives globally. Thus, it is clear that $\lambda_1 \leq \lambda_2$. We are interested in necessary and sufficient conditions to have $\lambda_1 < \lambda_2$. When the strict inequality holds we say that we have two phase transitions.

When there will be no ambiguity about the value of λ that we are considering we will drop λ from our notation.

Let X_t denote the continuous time Markov chain corresponding to $p(x, y)$. More precisely, consider a particle which undergoes random motion only (no branching). It waits a mean 1 exponential time and jumps from x to y with probability $p(x, y)$. We denote by X_t the position in S at time t of this particle. We define

$$P_t(x, y) = P(X_t = y | X_0 = x).$$

We will show below the following result.

$$\lim_{t \to \infty} \frac{1}{t} \log P_t(O, O) = -\gamma = \sup_{t > 0} \frac{1}{t} \log P_t(O, O).$$

Note that $\gamma \geq 0$ and since $P_t(O, O) \geq e^{-t}$ (if there are no jumps up to time t, the chain remains at O) we get that γ is in $[0,1]$.

We are now ready to state the main result of this section.

Theorem 1.1. *The first critical value for a branching Markov chain is*

$$\lambda_1 = 1.$$

The second critical value is

$$\lambda_2 = \frac{1}{1 - \gamma} \, \text{if } \gamma \text{ in } [0, 1)$$

$$\lambda_2 = \infty \, \text{if } \gamma = 1.$$

In particular there are two phase transitions for this model if and only if $\gamma \neq 0$.

Theorem 1.1 shows that the existence of two phase transitions is equivalent to the exponential decay of $P_t(O, O)$. This transforms a problem involving many particles into a problem involving only one particle.

The computation of λ_1 is easy and we do it now. Note that the process $|b_t^O|$ is a continuous time binary branching process. Each particle gives birth at rate λ and dies at rate 1. We have seen that this process survives with positive probability if and only if $\lambda > 1$. Hence, $\lambda_1 = 1$.

The computation of λ_2 is more involved and will be done below. We first give two applications of Theorem 1.1.

1.1   The Branching Random Walk on the Line

In this application we take $S = Z$ and

$$p(x, x + 1) = p \qquad p(x, x - 1) = q = 1 - p.$$

The system of particles evolves as follows. A particle at x gives birth to a particle at $x + 1$ at rate λp. A particle at x gives birth to a particle at $x - 1$ at rate λq. A particle dies at rate 1.

We now compute γ.

Lemma 1.1.

$$\lim_{t \to \infty} \frac{1}{t} \log P_t(O, O) = 2\sqrt{pq} - 1 = -\gamma.$$

Proof of Lemma 1.1. Since the continuous time Markov chain X_t with transition probabilities $P_t(x, y)$ waits a mean 1 exponential time between two jumps, the number of jumps up to time t is a Poisson process with parameter t. If we condition on the number of jumps up to time t we get:

$$P_t(x, y) = \sum_{n \geq 0} e^{-t} \frac{t^n}{n!} p_n(x, y)$$

where $p_n(x, y)$ is the probability that the discrete time chain goes from x to y in n steps.

Observe that $p_n(O, O) = 0$ if and only if n is odd. Therefore,

$$P_t(O, O) = \sum_{n \geq 0} e^{-t} \frac{t^{2n}}{(2n)!} p_{2n}(O, O).$$

We have seen that for a random walk on the line we have

$$\lim_{n \to \infty} \frac{p_{2n}(O, O)}{(4pq)^n (\pi n)^{-1/2}} = 1.$$

Hence, there are strictly positive constants $C_1 < 1$ and $C_2 > 1$ such that for all $n \geq 1$

$$C_1 (4pq)^n (\pi n)^{-1/2} \leq p_{2n}(O, O) \leq C_2 (4pq)^n (\pi n)^{-1/2}.$$

Therefore,

$$P_t(O, O) \leq e^{-t} + \sum_{n \geq 1} e^{-t} \frac{t^{2n}}{(2n)!} C_2 (4pq)^n (\pi n)^{-1/2}.$$

Note that

$$e^{-t} + \sum_{n \geq 1} e^{-t} \frac{t^{2n}}{(2n)!} C_2 (4pq)^n (\pi n)^{-1/2} \leq C_2 e^{t(-1+2\sqrt{pq})}.$$

Hence,

$$P_t(O, O) \leq C_2 e^{t(-1+2\sqrt{pq})}.$$

We now find a lower bound.

$$P_t(O, O) \geq e^{-t} + \sum_{n \geq 1} e^{-t} \frac{t^{2n}}{(2n)!} C_1 (4pq)^n (\pi n)^{-1/2}.$$

Set $C_3 = C_1/\sqrt{\pi}$ and observe that $\sqrt{n} \leq 2n + 1$ to get

$$P_t(O, O) \geq \frac{e^{-t}}{t} \sum_{n \geq 0} \frac{t^{2n+1}}{(2n+1)!} C_3 \sqrt{4pq}^{2n+1}.$$

But

$$\sum_{n \geq 0} \frac{t^{2n+1}}{(2n+1)!} = \frac{e^t - e^{-t}}{2}.$$

Thus, for t large enough we have (see the problems)

$$\sum_{n \geq 0} \frac{t^{2n+1}}{(2n+1)!} \geq e^t/4.$$

Hence,

$$P_t(O, O) \geq \frac{1}{4t} C_3 e^{t(-1+2\sqrt{pq})}.$$

We now put together the two inequalities to get, for t large enough,

$$\frac{1}{4t} C_3 e^{t(-1+2\sqrt{pq})} \leq P_t(O, O) \leq C_2 e^{t(-1+2\sqrt{pq})}.$$

We have

$$\lim_{t \to \infty} \frac{1}{t} \log(\frac{1}{4t} C_3 e^{t(-1+2\sqrt{pq})}) = \lim_{t \to \infty} \frac{1}{t} \log(C_2 e^{t(-1+2\sqrt{pq})}) = 2\sqrt{pq} - 1.$$

Thus,

$$\lim_{t \to \infty} \frac{1}{t} \log P_t(O, O) = 2\sqrt{pq} - 1$$

and this completes the proof of Lemma 1.1.

Now that γ has been computed we may apply Theorem 1.1 to get

$$\lambda_1 = 1 \qquad \lambda_2 = \frac{1}{2\sqrt{pq}}.$$

Observe that $pq \leq 1/4$ and that the equality holds if and only if $p = q = 1/2$. Therefore, the simple branching random walk on Z has two phase transitions if and only if $p \neq q$. In other words, any asymmetry in this model provokes the appearance of two phase transitions.

1.2 The Branching Random Walk on a Tree

The other application we will consider is the branching random walk on a homogeneous tree. Here S is a homogeneous tree (also called Bethe lattice) which is an infinite connected graph without cycles, in which every vertex has the same number of neighbors that we denote by $d \geq 3$, see Fig. 12.1.

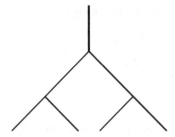

Fig. 12.1 This is part of the infinite binary tree ($d = 3$)

We assume here that $p(x, y) = 1/d$ for each of the d neighbors y of x. So in this sense this is a symmetric model, but we will see that the behavior is very similar to the one of the asymmetric branching random walk on Z.

Sawyer (1978) has computed asymptotic estimates of $p_n(O, O)$ for a large class of random walks on a homogeneous tree. In our particular case (nearest neighbor symmetric random walk) his computation gives

$$\lim_{n \to \infty} \frac{p_{2n}(O, O)}{n^{-3/2} R^{2n}} = C$$

where $C > 0$ is a constant and $R = \frac{2\sqrt{d-1}}{d}$. Doing the same type of computation as in Lemma 1.1 gives

$$\lim_{t \to \infty} \frac{1}{t} \log P_t(O, O) = R - 1 = \frac{2\sqrt{d-1}}{d} - 1 = -\gamma.$$

By Theorem 1.1

$$\lambda_1 = 1 \qquad \lambda_2 = \frac{d}{2\sqrt{d-1}}.$$

It is easy to check that for any $d \geq 3$,

$$\frac{d}{2\sqrt{d-1}} > 1$$

and therefore there are two distinct phase transitions for the simple symmetric branching random walk on any tree. For the symmetric simple random walk on the tree the probability that a walker goes back to the site he just left is $\frac{1}{d}$ while the probability he goes on to some other site is $\frac{d-1}{d}$. Hence, the walker is more likely to drift away rather than come back. In this sense this is similar to an asymmetric random walk on the line.

1.3 Proof of Theorem 1.1

We have already shown that $\lambda_1 = 1$. We now compute λ_2.
Our first step is to prove the existence of the limit

$$\lim_{t \to \infty} \frac{1}{t} \log P_t(O, O) = -\gamma = \sup_{t>0} \frac{1}{t} \log P_t(O, O).$$

Using the Markov property it is easy to see that

$$P_{t+s}(O, O) \geq P_t(O, O) P_s(O, O).$$

The existence of the limit γ is a direct consequence of the following lemma
applied to the function $f(t) = \log P_t(O, O)$.

Lemma 1.2. *If a continuous function f has the property that*

$$f(t + s) \geq f(t) + f(s)$$

then

$$\lim_{t \to \infty} \frac{1}{t} f(t) \text{ exists and is } = \sup_{t>0} \frac{1}{t} f(t) \in (-\infty, \infty].$$

Note that

$$P_t(x, y) = \sum_{n \geq 0} e^{-t} \frac{t^n}{n!} p_n(x, y).$$

This power series has an infinite radius of convergence (why?). Therefore, $f(t) = \log P_t(O, O)$ is continuous for all $t \geq 0$.

Proof of Lemma 1.2. Fix $s > 0$. Then for $t > s$ we can find an integer $k(t, s)$ such
that

$$0 \leq t - k(t, s)s < s.$$

Iterating the inequality $f(t + s) \geq f(t) + f(s)$ yields

$$f(t) \geq k(t, s) f(s) + f(t - k(t, s)s).$$

Let $m(s) = \inf_{0 < r < s} f(r)$. We get

$$\frac{1}{t} f(t) \geq \frac{1}{t} k(t, s) f(s) + \frac{1}{t} m(s).$$

We now let t go to infinity and use the fact that $k(t, s)/t$ converges to $1/s$ to get

$$\liminf_{t\to\infty} \frac{1}{t} f(t) \geq \frac{1}{s} f(s).$$

Hence

$$\liminf_{t\to\infty} \frac{1}{t} f(t) \geq \sup_{s>0} \frac{1}{s} f(s).$$

On the other hand, we have

$$\limsup_{t\to\infty} \frac{1}{t} f(t) \leq \sup_{s>0} \frac{1}{s} f(s).$$

This shows that

$$\liminf_{t\to\infty} \frac{1}{t} f(t) = \limsup_{t\to\infty} \frac{1}{t} f(t) = \sup_{s>0} \frac{1}{s} f(s).$$

This completes the proof of Lemma 1.2

Our second step in the proof of Theorem 1.1 is the following

Proposition 1.1. *For all x in S and for all times t we have the representation formula*

$$E(b_t^x(O)) = e^{(\lambda-1)t} P_{\lambda t}(x, O).$$

Proof of Proposition 1.1. We use the Kolmogorov backward differential equation for the random walk in continuous time. Conditioning on what happens in the time interval $[0, h]$ we have that

$$P_{t+h}(x, O) = \sum_{y\in S} hp(x, y) P_t(y, O) + (1 - h) P_t(x, O)$$

where we are neglecting terms of order higher than h. As $h \to 0$ we get

$$P_t'(x, O) = \sum_y p(x, y) P_t(y, O) - P_t(x, O). \tag{1.1}$$

Define $m(t, x) = E(b_t^x(O))$. We write again a backward equation. Conditioning on what happens in the interval $[0, h]$ and using that b_t^O is a Markov process gives

$$m(t+h, x) = \sum_{y\in S} \lambda hp(x, y)(m(t, x) + m(t, y)) + (1 - (\lambda+1)h) m(t, x). \tag{1.2}$$

Again we are neglecting terms of order higher than h. As $h \to 0$ in (1.2) we get

$$m'(t, x) = \sum_{y \in S} \lambda p(x, y) m(t, y) - m(t, x) \tag{1.3}$$

where the derivative is taken with respect to t. Equation (1.1) has a unique solution with the initial conditions $P_0(x, O) = 0$ for $x \neq O$ and $P_0(O, O) = 1$ (see Bhattacharya and Waymire 1990, Chap. IV, Sect. 3). This implies that

$$t \to e^{(\lambda-1)t} P_{\lambda t}(x, O)$$

is the unique solution to (1.3) with the initial value $m(0, x) = 0$ for $x \neq O$ and $m(0, O) = 1$.

This completes the proof of Proposition 1.1.

To prove Theorem 1.1 the crucial step is the following

Lemma 1.3. *If there is a time T such that $E(b_T^O(O)) > 1$, then*

$$\limsup_{t \to \infty} P(b_t^O(O) \geq 1) > 0.$$

Proof of Lemma 1.3. We will construct a supercritical Bienaymé–Galton–Watson process which is dominated (in a certain sense) by the branching Markov chain. To do so we will first consider a Markov process $\tilde{b}_t$ that is coupled to b_t^O in the following way. Up to time T $\tilde{b}_t$ and b_t^O are identical. At time T we suppress all the particles of $\tilde{b}_t$ which are not at site O and we keep the particles which are at O. Between times T and $2T$ the particles of $\tilde{b}_t$ which were at O at time T evolve like the particles of b_t^O which were at O at time T. At time $2T$ we suppress again all the particles of $\tilde{b}_t$ which are not at O. And so on, at times kT ($k \geq 1$) we suppress all the particles of $\tilde{b}_t$ which are not at O and between kT and $(k+1)T$ the particles of $\tilde{b}_t$ evolve like the corresponding particles of b_t^O.

Now we can define the following discrete time process Z_k. Let $Z_0 = 1$ and $Z_k = \tilde{b}_{kT}(O)$. We may write

$$Z_k = \sum_{i=1}^{Z_{k-1}} Y_i \text{ for } k \geq 1,$$

where Y_i is the number of particles located at O that a single particle initially located at O generates in T units time. In other words each Y_i has the same law as $b_T^O(O)$. Moreover the Y_i are independent one of the other and of the ones appearing in previous generations. Therefore Z_n is a Bienaymé–Galton–Watson process. By hypothesis, $E(Z_1) = E(b_T^O(O)) > 1$, so Z_k is a supercritical BGW process.

On the other hand, by our construction we have coupled the processes b_t^O and $\tilde{b}_t$ in such a way that $\tilde{b}_t(x) \leq b_t^O(x)$ for all x in S and all $t \geq 0$. Thus,

$$P(Z_k \geq 1) \leq P(b_{kT}^O(O) \geq 1).$$

But $P(Z_k \geq 1, \forall k \geq 0) > 0$ so making k go to infinity in the last inequality concludes the proof of Lemma 1.3.

We are finally ready to compute λ_2. Using

$$\lim_{t \to \infty} \frac{1}{t} \log P_t(O, O) = -\gamma = \sup_{t > 0} \frac{1}{t} \log P_t(O, O) \tag{1.4}$$

and Proposition 1.1 we get that for all $k \geq 0$

$$E(b_k^O(O)) \leq e^{Ck}$$

where $C = \lambda - 1 - \lambda\gamma$.

- Consider first the case $\gamma < 1$. Observe that if $\lambda < \frac{1}{1-\gamma}$ then $C < 0$. Let

$$A_k = \{b_k^O(O) \geq 1\}.$$

We have

$$P(A_k) \leq E(b_k^O(O)) \leq e^{Ck}.$$

By the Borel–Cantelli Lemma (see Sect. 2 of the Appendix),

$$P(\limsup_k A_k) = 0. \tag{1.5}$$

In other words $P(b_k^O(O) \geq 1$ for infinitely many $k) = 0$. So the process is dying out locally along integer times. We now take care of the noninteger times. If we had particles at O for arbitrarily large continuous times it would mean that all particles at O disappear between times k and $k + 1$, for infinitely many integers k. But for distinct k's these are independent events which are bounded below by a positive probability uniform in k. So the probability that this event happens for infinitely many k's is zero and (1.5) implies that

$$P(\limsup_{t \to \infty} b_t^O(O) \geq 1) = 0.$$

This shows that

$$\lambda_2 \geq \frac{1}{1 - \gamma}.$$

We will now prove the reverse inequality showing that λ_2 is actually equal to the r.h.s. Suppose that $\lambda > \frac{1}{1-\gamma}$. For $\epsilon > 0$ small enough we have that $\lambda > \frac{1}{1-\gamma-\epsilon}$. By (1.4) there is T large enough so that

$$\frac{1}{T} \log P_T(O, O) > -\gamma - \epsilon$$

and therefore by Proposition 1.1.

$$E(b_T^O(O)) \geq e^{DT}$$

with $D = \lambda(-\gamma - \epsilon + 1) - 1 > 0$. Since $E(b_T^O(O)) > 1$ we can apply Lemma 1.3 to get

$$\limsup_{t \to \infty} P(b_t^O(O) \geq 1) > 0. \tag{1.6}$$

We also have that

$$P(b_t^O(O) \geq 1) \leq P(\exists s \geq t : b_s^O(O) \geq 1).$$

We make $t \to \infty$ and get

$$\limsup_{t \to \infty} P(b_t^O(O) \geq 1) \leq P(\limsup_{t \to \infty} b_t^O(O) \geq 1)$$

this together with (1.6) shows that

$$P(\limsup_{t \to \infty} b_t^O(O) \geq 1) > 0.$$

Hence,

$$\lambda_2 \leq \frac{1}{1 - \gamma}.$$

This shows that if $\gamma < 1$ then

$$\lambda_2 = \frac{1}{1 - \gamma}.$$

• Consider now $\gamma = 1$. We use again that

$$P(A_k) \leq E(b_k^O(O)) \leq e^{Ck},$$

where $C = \lambda - 1 - \lambda \gamma = -1$ for any $\lambda > 0$. Hence, by Borel–Cantelli for any $\lambda > 0$

$$P(b_k^O(O) \geq 1 \text{ for infinitely many } k) = 0.$$

Therefore, $\lambda < \lambda_2$ for any $\lambda > 0$. This shows that $\lambda_2 = \infty$.
  This concludes the proof of Theorem 1.1.

Problems

1. Show that

$$\{\limsup_{t \to \infty} b_t^{O,\lambda}(O) \geq 1\} \subset \{|b_t^{O,\lambda}| \geq 1, \forall t > 0\}.$$

2. Show that $\lambda_1 \leq \lambda_2$.

3. Show that

$$e^{-t} + \sum_{n \geq 1} e^{-t} \frac{t^{2n}}{(2n)!} C_2 (4pq)^n (\pi n)^{-1/2} \leq C_2 e^{t(-1+2\sqrt{pq})}.$$

4. Let

$$-\gamma = \sup_{t>0} \frac{1}{t} \log P_t(O, O).$$

  Show that γ is in $[0, 1]$.

5. Show that for t large enough

$$\sum_{n \geq 0} \frac{t^{2n+1}}{(2n+1)!} \geq e^t/4.$$

6. Consider the simple branching random walk on Z^2. Show that any asymmetry provokes two phase transitions.

7. Consider an irreducible continuous time Markov chain on a finite graph with transition probabilities $P_t(x, y)$.

(a) Show that the corresponding γ is 0.
(b) How many phase transitions does a branching random walk on a finite graph have?

8. Show that $\gamma = 1$ if and only if for all $k \geq 1$ we have $p_k(O, O) = 0$. Use that

$$P_t(O, O) = \sum_{k \geq 0} e^{-t} \frac{t^k}{k!} p_k(O, O).$$

9. Prove that

$$P_{t+s}(O, O) \geq P_t(O, O) P_s(O, O).$$

10. Show that

$$P_t(x, y) = \sum_{n \geq 0} e^{-t} \frac{t^n}{n!} p_n(x, y)$$

is continuous for all $t \geq 0$.

11. To compute the critical values for the branching random walk on a tree we used that if

$$\lim_{n \to \infty} \frac{p_{2n}(O, O)}{n^{-3/2} R^{2n}} = C$$

then

$$\lim_{t \to \infty} \frac{1}{t} \log P_t(O, O) = R - 1.$$

Prove this implication. (Use a method similar to the one in Lemma 1.1.)

2 Continuity of the Phase Transitions

We will show that the first phase transition is continuous while the second one (when it exists) is not.

2.1 The First Phase Transition Is Continuous

Denote the global survival probability by

$$\rho(\lambda) = P(|b_t^{O,\lambda}| \geq 1, \forall t > 0).$$

Recall that the first critical value λ_1 has been defined as

$$\lambda_1 = \inf\{\lambda > 0 : \rho(\lambda) > 0\}.$$

Theorem 2.1. *The first phase transition is continuous in the sense that the function*

$$\lambda \to \rho(\lambda)$$

is continuous at λ_1.

Proof of Theorem 2.1. As noted before the total number of particles $|b_t^{O,\lambda}|$ is a continuous time branching process. We have shown that this allows to compute $\lambda_1 = 1$. Moreover, the critical process dies out. Thus, $\rho(\lambda_1) = 0$.

If $\lambda < \lambda_1$, then $\rho(\lambda) = 0$. So the limit from the left at λ_1 is

$$\lim_{\lambda \to \lambda_1^-} \rho(\lambda) = 0.$$

This together with $\rho(\lambda_1) = 0$ shows that ρ is left continuous at λ_1.

We now turn to the right continuity. This proof is very similar to a proof we did for percolation. We may simultaneously construct two branching Markov chains with birth rates, respectively, equal to $\lambda_1 p(x, y)$ and $\lambda_2 p(x, y)$ where $\lambda_1 < \lambda_2$. Denote the two processes by b_t^{O,λ_1} and b_t^{O,λ_2}. To do our simultaneous construction we construct b_t^{O,λ_2} in the usual way. That is, for all sites x, y in S each particle at x waits an exponential time with rate $\lambda_2 p(x, y)$ and gives birth to a new particle located at y. After each birth we consider a Bernoulli random variable independent of everything else which has a success probability equal to λ_1/λ_2. If we have a success, then a particle is also created at y for the process b_t^{O,λ_1} provided, that the particle at x which gives birth in the process b_t^{O,λ_2} also exists in the process b_t^{O,λ_1}.

This construction shows that the process with higher birth rates has more particles on each site of S at any time. In particular this implies that $\rho(\lambda)$ is increasing as a function of λ.

Consider now for a fixed time t the following function

$$f_t(\lambda) = P(|b_t^{O,\lambda}| \geq 1).$$

We will show that $f_t(\lambda)$ is continuous as a function of λ. By constructing the branching Markov chains with parameters λ and $\lambda + h$ we get for $h > 0$

$$0 \leq f_t(\lambda + h) - f_t(\lambda) = P(|b_t^{O,\lambda+h}| \geq 1; |b_t^{O,\lambda}| = 0). \tag{2.1}$$

Consider $N(t)$ the total number of particles born up to time t for the process $b_t^{O,\lambda+h}$. That is, we ignore the deaths and we count the births up to time t. From (2.1) we get for any positive integer n

$$0 \leq f_t(\lambda + h) - f_t(\lambda) \leq P(N(t) \leq n; |b_t^{O,\lambda+h}| \geq 1; |b_t^{O,\lambda}| = 0) + P(N(t) > n). \tag{2.2}$$

In order to have $|b_t^{O,\lambda+h}| > |b_t^{O,\lambda}|$, at least one of the Bernoulli random variables involved in the simultaneous construction must have failed. Therefore from (2.2) we get

$$0 \le f_t(\lambda + h) - f_t(\lambda) \le 1 - (\frac{\lambda}{\lambda + h})^n + P(N(t) > n).$$

We now make $h \to 0$ to get

$$0 \le \limsup_{h \to 0} f_t(\lambda + h) - f_t(\lambda) \le P(N(t) > n). \tag{2.3}$$

Note that $N(t)$ is a continuous time branching process. Each particle after an exponential time with parameter $a = \lambda + h$ is replaced by two particles with probability $f_2 = 1$. Hence,

$$E(N(t)) = e^{(\lambda+h)t}.$$

Observe that

$$P(N(t) > n) \le \frac{E(N(t))}{n}$$

see the problems. Note that, since t is fixed, the r.h.s. goes to 0 as n goes to infinity. Using the fact that the sequence $\{N(t) \ge n\}_{n \ge 0}$ is decreasing we have by Proposition 1.1 in the Appendix and the above inequality

$$\lim_{n \to \infty} P(N(t) > n) = P(\bigcap_{n \ge 1} \{N(t) > n\}) = P(N(t) = \infty) = 0. \tag{2.4}$$

Letting n go to infinity in (2.3) shows that $\lambda \to f_t(\lambda)$ is right continuous. The proof of left continuity is similar and we omit it. This proves that f_t is continuous.

Note that $\{|b_t^{O,\lambda}| \ge 1\} \subset \{|b_s^{O,\lambda}| \ge 1\}$ if $s < t$. By a continuous version of Proposition 1.1 in the Appendix we have

$$\lim_{t \to \infty} P(|b_t^{O,\lambda}| \ge 1) = P(|b_t^{O,\lambda}| \ge 1, \forall t \ge 0) = \rho(\lambda).$$

Moreover, since $f_t(\lambda) = P(|b_t^{O,\lambda}| \ge 1)$ is decreasing as a function of t we have

$$\rho(\lambda) = \inf_{t > 0} f_t(\lambda).$$

We proved (in the Percolation chapter) that the inf of continuous functions is upper semicontinuous and that a semicontinuous and increasing function is right continuous. This proves that ρ is right continuous and completes the proof of Theorem 2.1.

Remark. Observe that the proof that ρ is right continuous is fairly general and can be applied to a wide class of processes. That the limit from the left is zero at λ_1 is always true. So the main problem in showing that the phase transition is continuous is proving that $\rho(\lambda_1) = 0$. This is in general a difficult problem but here we take advantage of the branching structure of the process and there is no difficulty.

2.2   The Second Phase Transition Is Discontinuous

We say that the second phase transition is discontinuous in the following sense.

Theorem 2.2. *Assume that $p(x, y)$ is translation invariant. If the branching Markov chain has two distinct phase transitions, i.e. $\lambda_1 < \lambda_2$, then the function σ defined by*

$$\sigma(\lambda) = P(\limsup_{t \to \infty} b_t^{O,\lambda}(O) \geq 1)$$

is not continuous at λ_2.

Proof of Theorem 2.2. We will prove that if $\lambda > \lambda_2$ then

$$\sigma(\lambda) = P(\limsup_{t \to \infty} b_t^{O,\lambda}(O) \geq 1) = P(|b_t^{O,\lambda}| \geq 1, \text{ for all } t > 0) = \rho(\lambda). \quad (2.5)$$

In words, above λ_2 the process must survive locally if it survives globally. We now show that Theorem 2.2 follows directly from (2.5). Make λ approach λ_2 from the right in (2.5). Since ρ is continuous we have that $\rho(\lambda)$ approaches $\rho(\lambda_2)$. But this last quantity is strictly positive if $\lambda_1 < \lambda_2$ (why?). This shows that σ has a limit from the right at λ_2 which is strictly positive. But the limit from the left at λ_2 is zero (why?). So there is a discontinuity at λ_2.

We now turn to the proof of (2.5). For x, y in S we define B_x^y, the event that the site x is visited infinitely often by the offspring of a single particle started at the site y. That is,

$$B_x^y = \{\limsup_{t \to \infty} b_t^y(x) \geq 1\}.$$

Define

$$C^y = \bigcap_{x \in S} B_x^y.$$

In words, C^y is the event that the offspring of a particle initially at y visits all the sites infinitely often. Under the translation invariance assumptions for S and $p(x, y)$ we get $P(B_y^y) = P(B_O^O) = \sigma(\lambda) > 0$ if $\lambda > \lambda_2$. Observe that

$$|P(B_y^y) - P(B_x^y)| \leq P(B_y^y \cap (B_x^y)^c) + P(B_x^y \cap (B_y^y)^c).$$

In order for $B_y^y \cap (B_x^y)^c$ to happen it is necessary that y gets occupied at arbitrarily large times while after a finite time x is empty. But y is occupied at arbitrarily large times only if there are infinitely many distinct particles that occupy y after any finite time. Each of these particles has the same positive probability of occupying x and since distinct particles are independent of each other, x will get occupied with probability one at arbitrarily large times. Therefore $P(B_y^y \cap (B_x^y)^c) = 0$ and

$$P(B_y^y) = P(B_x^y) = \sigma(\lambda).$$

We now consider

$$P(B_x^y) - P(C^y) = P(B_x^y \cap (C^y)^c) \leq \sum_{z \in S} P(B_x^y \cap (B_z^y)^c).$$

For the same reason as above each term in this last sum is zero so

$$P(C^y) = P(B_x^y) = \sigma(\lambda) \tag{2.6}$$

for any $x, y \in S$. We have for any integer time k and integer n that

$$P(C^O) \geq P(C^O|\{|b_k^O| \geq n\}) P(|b_k^O| \geq n). \tag{2.7}$$

If the number of particles of b_k^O is m we denote by $B_k^O = \{y_1, y_2, \ldots, y_m\}$ an enumeration of the sites that are occupied by a particle at time k. Note that a site y will appear in B_k^O as many times as we have particles at y. We have

$$P(C^O|\{|b_k^O| \geq n\}) = \sum_{m \geq n} \sum_{y_1, \ldots, y_m} P(C^O; B_k^O = \{y_1, \ldots, y_m\}|\{|b_k^O| \geq n\}).$$

Note that if we have particles at sites $y_1, y_2, \ldots, y_m$ at time k, by the Markov property, in order to have C^O it is necessary and sufficient that one of the C^{y_i} occurs for $i = 1, 2, \ldots, m$. Thus,

$$P(C^O|\{|b_k^O| \geq n\}) = \sum_{m \geq n} \sum_{y_1, \ldots, y_m} P(\bigcup_{i=1}^m C^{y_i}) P(B_k^O = \{y_1, \ldots, y_m\}|\{|b_k^O| \geq n\}).$$

By the translation invariance of the model, we have $P(C^y) = P(C^O)$ for every y. Since offspring generated by different particles are independent we have for every sequence of sites $y_1, \ldots, y_m$

$$P(\bigcup_{i=1}^m C^{y_i}) = 1 - (1 - P(C^O))^m.$$

Since $m \geq n$, we get

$$P(C^O|\{|b_k^O| \geq n\}) \geq (1 - (1 - P(C^O))^n) \sum_{m \geq n} \sum_{y_1, \ldots, y_m} P(B_k^O = \{y_1, \ldots, y_m\}|\{|b_k^O| \geq n\}) =$$

$$1 - (1 - P(C^O))^n.$$

Using this lower bound in (2.7) gives

$$P(C^O) \geq (1 - (1 - P(C^O))^n) P(|b_k^O| \geq n). \tag{2.8}$$

Recall that $Z_k = |b_k^O|$ is a BGW. For a fixed $n \geq 1$, let

$$A_k = \{1 \leq Z_k < n\}.$$

By Proposition 1.2 in the Appendix

$$P(\limsup_k A_k) \geq \limsup_k P(A_k).$$

The event $\limsup_k A_k$ is the event that for infinitely many times k, Z_k is between 1 and n. This excludes the possibility that Z_k goes to zero or to infinity as k goes to infinity. But we know that a BGW either gets extinct or goes to ∞ as time goes to infinity. So, the event $\limsup_k A_k$ must have probability zero. Therefore,

$$\lim_{k \to \infty} P(A_k) = 0.$$

Observe that

$$P(A_k) = P(|b_k^O| \geq 1) - P(|b_k^O| \geq n),$$

so

$$\lim_{k \to \infty} P(|b_k^O| \geq n) = \lim_{k \to \infty} P(|b_k^O| \geq 1) = \rho(\lambda).$$

We use this observation and make k go to infinity in (2.8)

$$P(C^O) \geq (1 - (1 - P(C^O))^n)\rho(\lambda).$$

By (2.6) $P(C^O) = \sigma(\lambda) > 0$ for $\lambda > \lambda_2$, using this and making n go to infinity in the preceding inequality yields

$$\sigma(\lambda) \geq \rho(\lambda) \text{ for } \lambda > \lambda_2.$$

Since the reverse inequality is also true this concludes the proof of (2.5) and of Theorem 2.2.

Problems

1. Let X be a positive integer valued random variable with a finite expectation $E(X)$.

(a) Show that

$$P(X \geq n) \leq \frac{E(X)}{n}.$$

(b) Show that $P(X = \infty) = 0$.

2. Assume that $\lambda > \lambda_1$. Show that $\rho(\lambda) > 0$ where ρ is the probability of global survival.

3. Show that

$$\lim_{\lambda \to \lambda_2^-} \sigma(\lambda) = 0.$$

Notes

Branching random walks is one of the simplest spatial stochastic systems. It is possible to do explicit computations of critical values. As the next chapter illustrates this is rarely possible for other spatial stochastic processes. Theorem 1.1 was first proved in the particular case of trees by Madras and Schinazi (1992). The general case was proved by Schinazi (1993). Theorem 2.2 is due to Madras and Schinazi (1992). For more results on branching random walks see Cox (1994) and Liggett (1999).

References

Bhattacharya, R., Waymire, E.: Stochastic Processes with Applications. Wiley, New York (1990)
Cox, J.T.: On the ergodic theory of branching Markov chains. Stoch. Process. Appl. **50**, 1–20 (1994)
Liggett, T.M.: Stochastic Interacting Systems: Contact, Voter and Exclusion Processes. Springer, Heidelberg (1999)
Madras, N., Schinazi, R.B.: Branching random walks on trees. Stoch. Process. Appl. **42**, 255–267 (1992)
Sawyer, S.: Isotropic random walks in a tree. Z. Wahrsch. Verw. Gebiete **42**, 279–292 (1978)
Schinazi, R.B.: On multiple phase transitions for branching Markov chains. J. Stat. Phys. 71, 521–525 (1993)

Chapter 13
The Contact Process on a Homogeneous Tree

The contact process has the same birth and death rates as the branching random walk of the preceding chapter. The difference between the two models is that there is at most one particle per site for the contact process. The one particle per site condition makes offsprings of different particles dependent (unlike what happens for branching models). Exact computations become impossible. However, branching models are used to analyze the contact process.

1 The Two Phase Transitions

Let S be a homogeneous tree in which d branches emanate from each vertex of S. Thus, S is an infinite connected graph without cycles in which each site has d neighbors for some integer $d \geq 3$.

We consider the contact process on S whose state at time t is denoted by η_t. It evolves according to the following rules.

(i) If there is a particle at site $x \in S$, then for each of the d neighbors y of x it waits a mean $\frac{1}{\lambda}$ exponential time and then gives birth to a particle on y.

(ii) A particle waits a mean 1 exponential time and then dies.

(iii) There is at most one particle per site: births on occupied sites are suppressed.

The contact process follows the same rules as a branching Markov chain with the additional restriction (iii) that there is at most one particle per site for the contact process. This additional rule breaks the independence property between offspring of distinct particles that holds for branching Markov chains. Without independence we will not be able to make exact computations for the contact process. Instead, we will have to proceed by comparisons to simpler processes.

Let O be a distinguished vertex of the tree that we call the root. Let η_t^x be the contact process with only one particle at time 0 located at site $x \in S$. Let $\eta_t^x(y)$ be

© Springer Science+Business Media New York 2014

R.B. Schinazi, *Classical and Spatial Stochastic Processes: With Applications to Biology*, DOI 10.1007/978-1-4939-1869-0__13

the number of particles at site y and let $|\eta_t^x| = \sum_{y \in S} \eta_t^x(y)$ be the total number of particles. We define the critical values λ_1 and λ_2 corresponding to global and local survival, respectively.

$$\lambda_1 = \inf\{\lambda : P(|\eta_t^{O,\lambda}| \geq 1, \forall t > 0) > 0\}$$

$$\lambda_2 = \inf\{\lambda : P(\limsup_{t \to \infty} \eta_t^{O,\lambda}(O) = 1) > 0\}.$$

We include λ in the notation only when there may be an ambiguity about which value we are considering.

Our first result concerns bounds for the critical values.

Theorem 1.1. *We have the following bounds for the critical values.*

$$\frac{1}{d} \leq \lambda_1 \leq \frac{1}{d-2},$$

and

$$\lambda_2 \geq \frac{1}{2\sqrt{d-1}}.$$

In particular we have two phase transitions for the contact process (i.e. $\lambda_1 < \lambda_2$) on trees if $d \geq 7$.

Theorem 1.1 is a partial result. There are actually two phase transitions for any $d \geq 3$. However, the proof is rather involved. Our proof works only for $d \geq 7$ but is elementary.

Proof of Theorem 1.1. To get lower bounds for λ_1 and λ_2 we will consider a branching Markov chain that has more particles than the contact process. Define the branching Markov chain b_t^O where a particle at x gives birth to a particle at y with rate $\lambda d p(x,y)$, where $p(x,y) = 1/d$ if y is one of the d neighbors of x. A particle dies at rate 1. Since there is no restriction on the number of particles per site for b_t^O we may construct η_t^O and b_t^O simultaneously in such a way that $\eta_t^O(x) \leq b_t^O(x)$ for each x in S. We denote the two critical values of b_t by $\lambda_1(b)$ and $\lambda_2(b)$. Since b_t has more particles than η_t we have

$$\lambda_1 \geq \lambda_1(b) \text{ and } \lambda_2 \geq \lambda_2(b).$$

We have computed the critical values for this branching Markov chain. Observe that the parametrization is slightly different here and $d\lambda$ plays the role here that λ. So we get

$$d\lambda_1(b) = 1 \text{ and } d\lambda_2(b) = \frac{d}{2\sqrt{d-1}}.$$

This gives the lower bounds for λ_1 and λ_2 in Theorem 1.1.

To get an upper bound for λ_1, consider a process $\tilde{\eta}_t$ with the following rules. Start the process with a single particle at the root. Pick $d - 1$ sites among the d nearest neighbors. The particle at the root gives birth to a new particle at rate λ on each of the $d - 1$ sites previously picked. Each new particle can give birth on all neighboring sites but the parent site. Once a site has been occupied by a particle and this particle dies, the site remains empty forever. The death rate for each particle is 1 and there is at most one particle per site.

Define the distance between sites x and y in the homogeneous tree to be the length of the shortest path between x and y. Define $Z_0 = 1$ and Z_k to be the number of sites at distance k from O that will ever be occupied by a particle of $\tilde{\eta}_t$. Observe that each particle in $\tilde{\eta}_t$ gives birth (before dying) with probability $\frac{\lambda}{\lambda+1}$ on each of the $d - 1$ sites on which it is allowed to give birth. So the expected size of the offspring of each particle is

$$(d - 1)\frac{\lambda}{\lambda + 1}.$$

Since a tree has no cycles two distinct particles of $\tilde{\eta}_t$ have independent offspring. Hence Z_k is a BGW and it is supercritical if and only if

$$(d - 1)\frac{\lambda}{\lambda + 1} > 1.$$

The last inequality is equivalent to $\lambda > \frac{1}{d-2}$. On the other hand, it is clear that if the process Z_k survives if and only if the process $\tilde{\eta}_t$ survives. Hence, the first critical value of $\tilde{\eta}_t$ is $\frac{1}{d-2}$. Since the birth rules for $\tilde{\eta}_t$ are more restrictive than the one for η_t, we may construct η_t and $\tilde{\eta}_t$ simultaneously in such a way that $\tilde{\eta}_t(x) \leq \eta_t(x)$ for each x in S. This implies that $\frac{1}{d-2}$ is an upper bound for λ_1. This concludes the proof of Theorem 1.1.

Problems

1. Use the bounds in Theorem 1.1 to show that $\lambda_1 < \lambda_2$ if $d \geq 7$.

2. Show that the second critical value of the process $\tilde{\eta}_t$ (in the proof of Theorem 1.1) is infinite.

3. What are the critical values of a contact process on a finite graph?

2   Characterization of the First Phase Transition

While it is possible to improve the bounds in Theorem 1.1 (see the notes and references at the end of the chapter), the exact computation of the critical values is hopeless. In order to study the phase transitions we need to characterize the critical values in a way that is amenable to analysis. This is what we do next. Most results will not be proved, see the references for the proofs.

We start with the following.

Theorem 2.1. *For the contact process on a homogeneous tree with degree $d \geq 3$, there exist constants $c(\lambda, d)$ and $C(d)$ such that*

$$e^{c(\lambda,d)t} \leq E(|\eta_t^O|) \leq C(d)e^{c(\lambda,d)t}.$$

Moreover, $c(\lambda, d)$ is continuous as a function of λ.

Observe that is reminiscent of what happens for a branching process. In that case the expected number of particles is exactly an exponential function.

It is easy to prove the following.

Theorem 2.2. *If $\lambda > \lambda_1$, we have that $c(\lambda, d) > 0$.*

The following converse of Theorem 2.2 is much harder to prove.

Theorem 2.3. *If $c(\lambda, d) > 0$, then $\lambda \geq \lambda_1$.*

We can now state the characterization of the first phase transition.

Corollary 2.1. *We have that*

$$\lambda_1 = \sup\{\lambda : c(\lambda, d) \leq 0\}.$$

Moreover, $c(\lambda_1, d) = 0$.

Hence, λ_1 is the largest possible value for which we have $c(\lambda, d) = 0$.

Proof of Corollary 2.1. From Theorems 2.2 and 2.3 we get that

$$\lim_{\lambda \to \lambda_1^-} c(\lambda, d) \leq 0 \text{ and } \lim_{\lambda \to \lambda_1^+} c(\lambda, d) \geq 0.$$

Now using the fact that $\lambda \to c(\lambda, d)$ is a continuous function we get $c(\lambda_1, d) = 0$.

From Theorem 2.2 we know that λ_1 is an upper bound of the set $\{\lambda : c(\lambda, d) \leq 0\}$. We just saw that λ_1 is also in this set therefore

$$\lambda_1 = \sup\{\lambda : c(\lambda, d) \leq 0\}.$$

This completes the proof of Corollary 2.1.

An immediate consequence of Corollary 2.1 and Theorem 2.1 is

Corollary 2.2. *We have that at* $\lambda = \lambda_1$

$$1 \leq E(|\eta_t^{O,\lambda_1}|) \leq C(d)$$

where $C(d)$ *is a constant depending on d only.*

That is, the expected value of the number of particles of the critical contact process remains bounded at all times. This is similar to the critical branching process for which this expected value is a constant equal to one.

Corollary 2.3. *The survival probability for the contact process on a homogeneous tree with* $d \geq 3$

$$\lambda \rightarrow P(|\eta_t^{O,\lambda}| \geq 1, \forall t > 0)$$

is continuous at λ_1, *i.e., the first phase transition is continuous.*

It is also known that the first phase transition is continuous for the contact process on Z^d. For that model there is only one phase transition (i.e., $\lambda_1 = \lambda_2$). The question on Z^d was open for a long time and the proof is rather intricate (see Bezuidenhout and Grimmett 1990). In contrast, we are able to prove this result on the tree using elementary methods.

As for the second phase transition, the same type of argument that we used for Branching Markov chains works here too. The proof is complicated by the lack of independence in the contact process. See Madras and Schinazi (1992) for a proof of the following.

Theorem 2.4. *If* $\lambda_1 < \lambda_2$, *then the function*

$$\lambda \rightarrow P(\limsup_{t \to \infty} \eta_t^{O,\lambda}(O) = 1)$$

is not continuous at λ_2.

Problems

1. Show that Theorems 2.2 and 2.3 imply that

$$\lim_{\lambda \to \lambda_1^-} c(\lambda, d) \leq 0 \text{ and } \lim_{\lambda \to \lambda_1^+} c(\lambda, d) \geq 0.$$

2. Prove Corollary 2.2.

3. Show that Theorem 2.2 implies that λ_1 is an upper bound of the set

$$\{\lambda : c(\lambda, d) \leq 0\}.$$

Notes

The contact process is an example of an *interacting particle system*. These are spatial stochastic models for which space is discrete and time is continuous. The analysis of these models requires advanced mathematics. Liggett (1985, 1999) has provided excellent accounts of the progress in this field.

The contact process (on Z^d) was first introduced by Harris (1974). Pemantle (1992) started the study of the contact process on trees and proved that there are two phase transitions for all $d \geq 4$. Liggett (1996) and Stacey (1996) have independently proved that there are two phase transitions for $d = 3$ as well.

The proofs that are not provided here can be found Madras and Schinazi (1992) and Morrow et al. (1994).

References

Bezuidenhout, C., Grimmett, G.: The critical contact process dies out. Ann. Probab. **18**, 1462–1482 (1990)

Harris, T.E.: Contact interactions on a lattice. Ann. Probab. **6**, 198–206 (1974)

Liggett, T.M.: Interacting Particle Systems. Springer, New York (1985)

Liggett, T.M.: Multiple transition points for the contact process on the binary tree. Ann. Probab. **26**, 1675–1710 (1996)

Liggett, T.M.: Stochastic Interacting Systems: Contact, Voter and Exclusion Processes. Springer, Heidelberg (1999)

Madras, N., Schinazi, R.: Branching random walks on trees. Stoch. Process. Appl. **42**, 255–267 (1992)

Morrow, G., Schinazi, R., Zhang, Y.: The critical contact process on a homogeneous tree. J. Appl. Probab. **31**, 250–255 (1994)

Pemantle, R.: The contact process on trees. Ann. Probab. **20**, 2089–2116 (1992)

Stacey, A.: The existence of an intermediate phase for the contact process on trees. Ann. Probab. **24**, 1711–1726 (1996)

Appendix A
A Little More Probability

In this appendix we review a few more advanced probability tools. These are useful in the analysis of the spatial models.

1 Probability Space

Consider Ω a countable space. A probability P is a function from the set of subsets of Ω to $[0,1]$ with the two following properties.

$$P(\Omega) = 1$$

and if $A_n \subset \Omega$ for $n \geq 1$, and $A_i \cap A_j = \emptyset$ for $i \neq j$ then

$$P(\bigcup_{n\geq 1} A_n) = \sum_{n\geq 1} P(A_n).$$

The subsets of Ω are called events. We say that the sequence of events A_n is increasing if $A_n \subset A_{n+1}$ for $n \geq 1$. The sequence A_n is said to be decreasing if $A_{n+1} \subset A_n$ for $n \geq 1$.

Proposition 1.1. *Let (Ω, P) be a probability space, A, B, and A_n be events. We have the following properties:*

(i) If $B \subset A$, then $P(A \cap B^c) = P(A) - P(B)$.
(ii) For any sequence of events A_n we have $P(\bigcup_{n\geq 1} A_n) \leq \sum_{n\geq 1} P(A_n)$.
(iii) If A_n is a sequence of increasing events, then $\lim_{n\to\infty} P(A_n) = P(\bigcup_{n\geq 1} A_n)$.
(iv) If A_n is a sequence of decreasing events, then $\lim_{n\to\infty} P(A_n) = P(\bigcap_{n\geq 1} A_n)$.

© Springer Science+Business Media New York 2014
R.B. Schinazi, *Classical and Spatial Stochastic Processes: With Applications to Biology*, DOI 10.1007/978-1-4939-1869-0

Proof of Proposition 1.1. Observe that B and $A \cap B^c$ are disjoint and their union is A. Hence

$$P(A) = P(A \cap B^c) + P(B)$$

and this proves (i).

To prove (iii) assume A_n is an increasing sequence of events. Define

$$A = \bigcup_{n \geq 1} A_n \text{ and } B_1 = A_1, B_n = A_n \cap A_{n-1}^c \text{ for } n \geq 2.$$

The B_n are disjoint and their union is still A (why?) therefore

$$P(A) = \sum_{n \geq 1} P(B_n) = P(A_1) + \lim_{n \to \infty} \sum_{p=2}^{n} (P(A_p) - P(A_{p-1})).$$

So we get

$$P(A) = P(A_1) + \lim_{n \to \infty} (P(A_n) - P(A_1))$$

and this proves (iii).

We now use (iii) to prove (ii). For any sequence of events A_n we may define

$$C_n = \bigcup_{p=1}^{n} A_p.$$

C_n is increasing. We also have that for any two events A and B

$$P(A \cup B) = P((A \cap (A \cap B)^c) \cup B) = P(A \cap (A \cap B)^c) + P(B).$$

Since

$$P(A \cap (A \cap B)^c) = P(A) - P(A \cap B)$$

we have

$$P(A \cup B) = P(A) + P(B) - P(A \cap B).$$

Hence,

$$P(A \cup B) \leq P(A) + P(B)$$

and by induction we get for any finite union

$$P(C_n) = P(\bigcup_{p=1}^{n} A_p) \leq \sum_{p=1}^{n} P(A_p). \tag{1.1}$$

Using (iii) we know that $P(C_n)$ converges to the probability of the union of C_n which is the same as the union of the A_n. Making n go to infinity in (1.1) yields

$$\lim_{n \to \infty} P(C_n) = P(\bigcup_{p=1}^{\infty} A_p) \leq \sum_{p=1}^{\infty} P(A_p).$$

This concludes the proof of (ii).

For (iv) it is enough to observe that if A_n is a decreasing sequence then A_n^c is an increasing sequence and by (iii)

$$\lim_{n \to \infty} P(A_n^c) = P(\bigcup_{n \geq 1} A_n^c) = P((\bigcap_{n \geq 1} A_n)^c) = 1 - P(\bigcap_{n \geq 1} A_n)$$

and this concludes the proof of Proposition 1.1.

Let $(a_n)_{n \geq 1}$ be a sequence of real numbers. Observe that $b_n = \sup_{p \geq n} a_p$ defines a decreasing sequence (if the supremum does not exist we take $b_n = \infty$). Then $\lim_{n \to \infty} b_n = \inf_{n \geq 1} b_n$ exists (it may be finite or infinite) and we denote

$$\limsup a_n = \lim_{n \to \infty} b_n = \inf_{n \geq 1} \sup_{p \geq n} a_p.$$

In a similar way one can define

$$\liminf a_n = \sup_{n \geq 1} \inf_{p \geq n} a_p.$$

The next result gives the relation between limits and lim sup and lim inf.

Theorem 1.1. *The limit of a_n exists if and only if* $\liminf a_n = \limsup a_n$ *and in that case we have*

$$\lim_{n \to \infty} a_n = \liminf a_n = \limsup a_n.$$

For a proof and other properties of sequences see an analysis text such as Rudin (1976).

We will also need the following property: if $(a_n)_{n \geq 1}$ and $(c_n)_{n \geq 1}$ are two sequences of real numbers such that for each $n \geq 1$, $a_n \leq c_n$, then

$$\liminf a_n \leq \liminf c_n.$$

By analogy with the real numbers, for any sequence of events A_n we define the events

$$\limsup A_n = \bigcap_{n \geq 1} \bigcup_{p \geq n} A_p \text{ and } \liminf A_n = \bigcup_{n \geq 1} \bigcap_{p \geq n} A_p.$$

Proposition 1.2. *We have the following inequalities for any sequence of events A_n*

$$P(\liminf A_n) \leq \liminf P(A_n)$$

$$P(\limsup A_n) \geq \limsup P(A_n).$$

Proof of Proposition 1.2. Define $B_n = \bigcap_{p \geq n} A_p$ and observe that B_n is an increasing sequence of events. Hence by Proposition 1.1

$$\lim_{n \to \infty} P(B_n) = P\left(\bigcup_{n \geq 1} B_n\right) = P(\liminf A_n).$$

Since $P(A_n) \geq P(B_n)$ we get

$$\liminf P(A_n) \geq \lim_{n \to \infty} P(B_n) = P(\liminf A_n)$$

and this proves the first inequality in Proposition 1.2. The second inequality is left to the reader.

Problems

1. Prove that $P(\emptyset) = 0$ and that $P(A^c) = 1 - P(A)$ for any subset A of Ω.

2. Show that for any events A and B

$$P(A) = P(A \cap B) + P(A \cap B^c).$$

3. Prove that for any events A and B we have

$$|P(A) - P(B)| \leq P(A \cap B^c) + P(A^c \cap B).$$

4. Check that ω is in $\limsup A_n$ if and only if ω is in A_n for infinitely many distinct n. Check that ω is in $\liminf A_n$ if and only if ω is in A_n for all n except possibly for a finite number of n.

5. Prove the second inequality in Proposition 1.2.

2   Borel–Cantelli Lemma

Given an event B such that $P(B) > 0$ we define the conditional probability

$$P(A|B) = \frac{P(A \cap B)}{P(B)}.$$

We say that the events A and B are independent if $P(A \cap B) = P(A)P(B)$. More generally we say that the events $A_1, A_2, \ldots, A_n$ are independent if for all integers $i_1, i_2, \ldots, i_p$ in $\{1, \ldots, n\}$ we have

$$P(\bigcap_{j=1}^{p} A_{i_j}) = \Pi_{j=1}^{p} P(A_{i_j}).$$

We now state a very useful property.

Borel-Cantelli Lemma *If A_n is a sequence of events such that*

$$\sum_{n \geq 1} P(A_n) < \infty$$

then $P(\limsup A_n) = 0$.
Conversely, if the A_n are independent events and

$$\sum_{n \geq 1} P(A_n) = \infty$$

then $P(\limsup A_n) = 1$.
Observe that the independence assumption is only needed for the converse.

Proof of the Borel–Cantelli Lemma We first assume that $\sum_{n \geq 1} P(A_n) < \infty$.
Define $B_n = \bigcup_{p \geq n} A_p$. Since B_n is a decreasing sequence we have by Proposition 1.1

$$\lim_{n \to \infty} P(B_n) = P(\bigcap_{n \geq 1} B_n) = P(\limsup A_n).$$

On the other hand

$$P(B_n) \leq \sum_{p \geq n} P(A_p),$$

but the last sum is the tail of a convergent series, therefore

$$\lim_{n\to\infty} P(B_n) = P(\limsup A_n) \le \lim_{n\to\infty} \sum_{p\ge n} P(A_p) = 0.$$

For the converse we need the two assumptions. The A_n are independent and the series is infinite. Using the fact that the A_n^c are independent, for any integers $m < n$ we have that

$$P\left(\bigcap_{p=m}^{n} A_p^c\right) = \Pi_{p=m}^{n} P(A_p^c) = \Pi_{p=m}^{n}(1 - P(A_p)).$$

Since $1 - u \le e^{-u}$ we get

$$P\left(\bigcap_{p=m}^{n} A_p^c\right) \le e^{-\sum_{p=m}^{n} P(A_p)}. \tag{2.1}$$

Fix m and define $C_n = \bigcap_{p=m}^{n} A_p^c$. The sequence C_n is decreasing and

$$\bigcap_{n\ge m} C_n = \bigcap_{p=m}^{\infty} A_p^c.$$

We now make n go to infinity in (2.1) to get, by Proposition 1.1,

$$\lim_{n\to\infty} P(C_n) = P\left(\bigcap_{n\ge m} C_n\right) = P\left(\bigcap_{p=m}^{\infty} A_p^c\right) \le \lim_{n\to\infty} e^{-\sum_{p=m}^{n} P(A_p)} = 0$$

where we are using that for any m,

$$\lim_{n\to\infty} \sum_{p=m}^{n} P(A_p) = \infty.$$

So we have that

$$P\left(\left(\bigcap_{p=m}^{\infty} A_p^c\right)^c\right) = P\left(\bigcup_{p=m}^{\infty} A_p\right) = 1 \text{ for every } m \ge 1.$$

Since $D_m = \bigcup_{p=m}^{\infty} A_p$ is a decreasing sequence of events, we let m go to infinity to get

$$P\left(\bigcap_{m\ge 1} \bigcup_{p=m}^{\infty} A_p\right) = P(\limsup A_n) = 1.$$

This completes the proof of the Borel–Cantelli Lemma.

2.1   Infinite Products

We use Borel–Cantelli Lemma to prove useful results on infinite products.

Consider a sequence s_j in $(0,1)$. Define $P_n = \Pi_{j=0}^n (1 - s_j)$. P_n is a decreasing sequence (why?) and it is bounded below by 0. Therefore P_n converges. We define the following infinite product as the limit of P_n.

$$P = \Pi_{j=0}^\infty (1 - s_j) = \lim_{n \to \infty} P_n.$$

Note that for all $n \geq 0$ we have $0 < P_n < 1$ and so $0 \leq P \leq 1$. We would like to know when $P = 0$. We will show that

$$P = \Pi_{j=0}^\infty (1 - s_j) > 0 \text{ if and only if } \sum_{j=0}^\infty s_j < \infty.$$

We now introduce some probability in order to prove the claim. Consider a sequence of independent random variables $(X_n)_{n \geq 0}$ such that $P(X_n = 0) = 1 - s_n$ and $P(X_n = 1) = s_n$. Let

$$A_n = \{X_n = 0\} \text{ and } B_n = \bigcap_{j=0}^n A_j.$$

Note that

$$P_n = \Pi_{j=0}^n (1 - s_j) = P(B_n).$$

Moreover B_n is a decreasing sequence of events and so by letting n go to infinity in the preceding equality we get by Proposition 1.1

$$P = \Pi_{j=0}^\infty (1 - s_j) = P(\bigcap_{j=0}^\infty B_j).$$

Note that

$$\{\bigcap_{j=0}^\infty B_j\} = \{\bigcap_{j=0}^\infty A_j\}.$$

Hence, P is the probability that all X_n are 0.

Assume first that the series $\sum_{j=0}^\infty s_j$ diverges. Thus,

$$\sum_{n \geq 0} P(A_n^c) = \infty$$

and since the events A_n are independent we may conclude (by Borel–Cantelli) that

$$P(\limsup A_n^c) = 1.$$

That is, with probability one there are infinitely many $X_n = 1$. So P (the probability that all $X_n = 0$) must be 0. This proves half of the claim.

We now assume that the series $\sum_{j=0}^{\infty} s_j$ converges. By definition of convergence, we have that for k large enough

$$\sum_{j=0}^{\infty} s_j - \sum_{j=0}^{k-1} s_j = \sum_{j \geq k} s_j < 1.$$

Fix k so that the inequality above holds. Now consider the inequality

$$P\left(\bigcup_{j \geq k} A_j^c\right) \leq \sum_{j \geq k} P(A_j^c) = \sum_{j \geq k} s_j < 1.$$

In other words, this shows that the probability of having at least one $X_j = 1$ for $j \geq k$ is strictly less than 1. Therefore, we have

$$1 - P\left(\bigcup_{j \geq k} A_j^c\right) = P\left(\bigcap_{j \geq k} A_j\right) > 0.$$

We now write the infinite product as a finite product:

$$P = P(\cap_{j \geq 0} A_j) = P(A_0)P(A_1)\dots P(A_{k-1})P(\cap_{j \geq k} A_j).$$

Since each factor in the finite product on the right hand-side is strictly positive, P is also strictly positive and we are done.

Problems

1. Prove that if A and B are independent so are A and B^c, and A^c and B^c.

2. A coin is tossed until it lands on heads. The probability that the coin lands on heads is p. Let X be the number of tosses to get the first heads.

(a) Show that

$$P(X = k) = (1 - p)^{k-1} p \text{ for } k = 1, 2, \dots.$$

(b) Show that for $k \geq 1$

$$P(X > k) = (1 - p)^k.$$

(c) Consider now a sequence X_n, $n \geq 1$, of independent identically distributed random variables, each with the distribution above. Show that

$$(1 - p)n^{ac} \leq P(X_n > a \ln n) \leq n^{ac},$$

where $c = \ln(1 - p)$ and $a > 0$.

(d) Show that if $ac < -1$ then with probability one there are only finitely many n for which $X_n > a \ln n$.

(e) Show that if $ac \geq -1$ there are infinitely many n such that $X_n > a \ln n$.

Notes

Feller (1968) is a great reference for probability theory on countable spaces. At a more advanced level (using measure theory) there are the books of Durrett (2010) and Port (1994).

References

Durrett, R.: Probability: Theory and Examples, 4th edn. Cambridge University Press, Cambridge (2010)

Feller, W.: An Introduction to Probability Theory and its Applications, vol. I, 3rd edn. Wiley, New York (1968)

Port, S.C.: Theoretical Probability for Applications. Wiley, New York (1994)

Rudin, W.: Principles of Mathematical Analysis, 3rd edn. McGraw-Hill, New York (1976)

Index

Printed in the United States
By Bookmasters